SpringerBriefs in Ethics

More information about this series at http://www.springer.com/series/10184

Aviad E. Raz · Silke Schicktanz

Comparative Empirical Bioethics: Dilemmas of Genetic Testing and Euthanasia in Israel and Germany

Springer

Aviad E. Raz
Department of Sociology and Anthropology
Ben-Gurion University of the Negev
Beersheba
Israel

Silke Schicktanz
University Medical Center
University of Göttingen
Göttingen
Germany

ISSN 2211-8101 ISSN 2211-811X (electronic)
SpringerBriefs in Ethics
ISBN 978-3-319-32731-0 ISBN 978-3-319-32733-4 (eBook)
DOI 10.1007/978-3-319-32733-4

Library of Congress Control Number: 2016936996

Printed on acid-free paper

This Springer imprint is published by Springer Nature
The registered company is Springer International Publishing AG Switzerland

Acknowledgments

This book would not have been written without all the lay participants taking part in interviews and focus groups discussions in Germany and Israel during 2005–2007 and 2009–2011. We would like to thank all the respondents who participated in this study. Before and during our joint data collection we were also involved in cross-cultural networking that paved the way for the work on which this book is based. We co-organized and participated in a series of three German-Israeli expert workshops. The first workshop in this series was "The Influence of Religion and Culture on Biomedical Sciences: A German-Israeli Dialogue about Medical Possibilities at the Edges of Life" held in November 2007 in Berlin. The second workshop was titled "Genetics and Society: public/professional and discourse/practice" and held in BGU in December 2008. The work presented in this workshop was published in a special issue (co-edited by us) of the journal *New Genetics and Society*. The third workshop in this series was a German-Israeli "winter school" on "The culture and ethics of biomedicine" held at the University Medical Center of Göttingen in 2009. All these workshops helped to refine and reinvigorate our collaboration, and we would like to express our debt and gratitude to the BMBF and to our Universities that provided funding for these activities. During this period we compared and discussed data collected separately by us in the context of lay attitudes toward genetic testing (Raz and Schicktanz 2009a, b).

Our joint research project, on which this book is based, was made possible thanks to the generous support of the German-Israeli Foundation for Science and Development: Young scientist GIF Grant No. *G-21071451-4/2005* "Bodies in Culture—Bioethical aspects of biomedicine and its socio-cultural contexts," and GIF Grant No. *1023–317.4/2008* "Cross-Cultural Ethics of Health and Responsibility: Expert and lay perspectives regarding bioethical dilemmas in Germany and Israel." Needless to say, we are extremely grateful to the GIF for this invaluable support. Moreover, we would like to thank all the German and Israeli lay persons, patients, and professionals who shared their experiences and insights with us during our empirical research.

We are very grateful to the enormous help of our dedicated team of young research assistants: Dr. Nitzan Rimon-Zarfaty, Dr. Julia Inthorn and Dr. Mark Schweda. Nitzan had an important role in conducting, coding, and co-analyzing the focus groups conducted in 2009–2011 in Israel. She was also intensively involved in developing the coding scheme and peer-coding. Julia Inthorn conducted, coded, and co-analyzed the German focus groups in 2010–2011. She was also intensively involved in developing the coding scheme and peer-coding. Mark Schweda was involved in preparing the first phase of focus groups in 2009 as well as in the theoretical conceptualization of the meanings of "being affected" and "responsibility." We are very grateful for the important contributions made by these three scholars and we are very proud to see them making progress on their independent paths.

Aviad Raz and Silke Schicktanz were the PIs who jointly conceptualized and wrote the research proposals, conceptualized and substantially led the final analysis and structure of all the papers which are used in this book. Parts of Chapter 5 have greatly benefitted from suggestions made by Barbara Prainsack, Gabriele Werner-Felmayer, Yael Hashiloni-Dolev and Carmel Shalev. Felix Hagenström's support was of great help in copy editing the manuscript.

Writing this book was a long process that took place in most of 2014 and 2015. We reviewed the data produced in our previous studies, reflecting on the common denominators and emerging insights that we were now aware of in retrospect. The book endeavors to locate this project in a broader context, presenting what is hopefully a more clear and general crystallization of our work on socio-empirical bioethics and the "Israeli" and "German," "religious" and "secular," "affected" and "non-affected" narratives of responsibility. Large portions from the following papers are reprinted in this book by permission of the publishers: Schicktanz S, Schweda M (2012) The diversity of responsibility: The value of explication and pluralization. Medicine Studies 3:131–145; Raz A, Rimon-Zarfaty N, Inthorn J, Schicktanz S (2014) Making responsible life plans: Cultural differences in lay attitudes toward predictive genetic testing for late-onset diseases. In: Prainsack B, Schicktanz S, Werner-Felmeyer G (eds) Genetics as social practice. Ashgate, Farnham, pp 181–198; Schicktanz S, Raz A, Shalev C (2010) The cultural context of patient's autonomy and doctor's duty: Passive euthanasia and advance directives in Germany and Israel. Medicine, Health Care and Philosophy 13(4):363–369.

Silke Schicktanz thanks the Alfried Krupp Wissenschaftskolleg Greifswald for providing her with a fellowship from April to September 2015. The special working atmosphere allowed her to finalize her part of this book. Aviad Raz wishes to express his gratitude to the College of Management Academic Studies, Rishon Lezion, Israel, for the financial support provided for this publication.

The authors' names appear in alphabetical order; however, the contents of this book are our mutual responsibility.

Omer
Göttingen/Greifswald
2015

Aviad E. Raz
Silke Schicktanz

Contents

Chapter 1
Introduction: Engaging in Comparative Bioethics

In 2004, an Israeli sociologist and a German bioethicist met at an international conference on bioethics in a cloister in the Netherlands. Following a presentation they started conversing and the Israeli sociologist asked: "How is it that Germans are so reluctant about genetic testing?" The German bioethicist paused a bit and answered with uncertainty: "I am not sure whether they are so reluctant. The geneticists I met are all very supportive." After sharing some comparative data, they realized that the existing research consisted mainly of policy analysis and surveys of professionals. There were two important lacunas. First, public opinion and lay attitudes have not been well-studied. Second, the question "why" was still open, requiring a more empirically grounded and nuanced understating that ventures beyond important (yet easily flagged) premises such as the historical legacy of the Holocaust—an issue that is perhaps always in the background when a German and an Israeli meet.

Indeed, our task in this book concerning a contemporary re-examination of the lessons of the Holocaust presents a relatively new stage in the history of German-Israeli research collaborations in the social sciences and the humanities. The first stage (around the 1950s–1960s) could be characterized by avoidance, with some beginnings of German-Israeli collaborations in the basic sciences but none in the humanities. In the next stage (around the 1970s–1980s), the subject of "German history" was allowed back into Israeli universities, nevertheless while avoiding the Holocaust. This book is the outcome of a relatively recent phase of re-examination, expressed by studies that take the Holocaust as a focus for re-examining present-day cultural scripts of eugenics and euthanasia with a focus on the common, yet also very different, legacy and interpretation of the lessons of the Holocaust.

Five months after their first and unplanned meeting, the German bioethicist and the Israeli sociologist embarked on an intensive exchange by making plans for joint research projects. Now, ten years later, we decided to write a book. This book is a comprehensive, empirically-grounded exploration of a culturally embedded bioethics. Its aim is to develop, methodologically and theoretically, a research

A.E. Raz and S. Schicktanz, *Comparative Empirical Bioethics: Dilemmas of Genetic Testing and Euthanasia in Israel and Germany*, SpringerBriefs in Ethics, DOI 10.1007/978-3-319-32733-4_1

program that engages comparative bioethics and sociology, bringing together socio-empirical methods, ethical and cultural theories, and new meta-concepts. When we use here the term "bioethics," we refer to the (relatively new, but quickly spreading) analysis of how "bioethical" issues become everyday questions that are imbued with meanings by lay people in their life, as they are confronted with modern medicine and have to make decisions about life and death. In the last decade, various scholars have attempted to broaden our view of "bioethics" at least in two new and inter-connected directions: the cultural embedding of bioethics, and a criticism of the regime of experts. Both directions are very important for this book.

1.1 One World, One Bioethics?

When scholars started comparing bioethical discourses across nations, they showed intriguingly how cloning, stem cell research, or end-of-life issues are discussed differently by ethical experts in different countries (Bleiklie et al. 2004; Blank and Merrick 2007; Jasanoff 2007). However, many differences that were found in this earlier stage of analysis were the result of different legal frameworks. Liberal bioethics tends to focus in narrow terms on the legal-political question whether a particular practice should be forbidden or left unregulated (meaning that it is actually allowed). What this first cross-national comparison revealed for some Western bioethicists, was that the normative priority given to individual autonomy and self-determination was not so globally accepted as it seemed within an Anglo-American discourse (Tao 2002). The "standard code of Western bioethics" turned out to be not so standard. Although the discussion about moral universalism (and global bioethics) versus moral particularism is already well-known, it has not provided a practical way to work through this dichotomy. It was time to move beyond the classical question whether there are universally valid and applicable norms of global bioethics, such as, "the right to live," "respect for autonomy" or "human dignity," or whether norms are only valid and justified in a particular setting of place and time (e.g., the discussion of "Asian versus Western bioethics"; De Castro 1999). Both options, it appeared, had their internal logic and consistency.

According to our understanding, this debate has so far been neither sufficient nor very fruitful. The reasons for the differences found were often not sufficiently understood. Instead of considering seriously the rationales for socio-ethical diversity, and critically reflecting whether there were consistent, homogenous positions in one context or rather diverse voices, academic ethicists are occupied by defining new avenues for a universal bioethics. We are not against the idea of global bioethical principles. Given the globalization of medicine and its manifold implications, this is a valuable, even inevitable goal. However, according to our understanding we are just scratching the surface of the complex picture. If we want to see what is behind the first, declarative layer, we need in-depth understanding of the cultural embeddings by bringing forward lay moralities (rather than merely policies and professional views) and acknowledging their intra-contextual diversity.

The foundation of a convincing, socially robust global bioethics requires the recognition of differences and existing varieties.

In the broad spectrum of approaches to bioethics, three major streams can be differentiated. Even though we refer here mainly to dominant Western authors, the different approaches can also be found within the Asian, African or South-American debate, but often with different key authors[1]:

First, there is the deductive, abstract approach from general ethical theory to a situation-specific normative orientation. This approach, often analytically clear and theory-driven, is still common when scholars refer to Peter Singer's utilitarianism, Immanuel Kant's deontology or Aristotle's virtue ethics (e.g., Gauthier 2002), for instance when assessing the practical questions in medical ethics (Singer 2003; Tassicker et al. 2003; Candee and Puka 1984). Secondly, and on the other end of the spectrum, we find inductive, biographical narratives that serve as moral justification in casuistic, case-by-case-assessment. In this approach, each and every person is seen as capable of moral reasoning. These two approaches can be criticized: the former may be seen as too rigid, prefixed, western-centric and theoretically one-sided, while the latter can be regarded as advocating normative relativism (Brandt 1961, pp. 433–440; Bruchhausen 2003; Cook 1999). This dispute will likely go on forever.[2] Recent approaches coming from feminist ethics, narrative bioethics, or pragmatism cannot really escape this dichotomy, as the discussion of the relational concept of autonomy reveals (Mackenzie and Stoljar 2000).

The third, prominent, so-called middle-range principle approach, advocated by the two North-American bioethicists Tom Beauchamp and James Childress (Beauchamp and Childress 2013), suggests to use four commonly shared principles "respect for autonomy, beneficence, maleficence and justice" as moral orientation in concrete health care issues. These principles were developed to solve practical moral conflicts, ubiquitous in health care. However, this popular path has somehow developed into a dogma or ideology. Insisting on the infallibility and exclusivity of these four principles may conflict with the complexity and diversity of health-care situations.[3] Furthermore, one needs to rely again on different, often contrary ethical theories when situating each principle in an applied context: Whether one conceptualizes autonomy from Kantian deontology or from liberal consequentialism

[1]We do not list here exhaustively all these authors as they are not globally so important as it might look like.

[2]While this is often to be understood as a destructive process, it can be also seen as a rather very productive procedure of scientific practice as many sub-approaches, logic critiques and pitfalls are detected on both sides.

[3]A more productive, but less ideological application of this approach is to see the four principles as "heuristic matrix" for the identification of major moral conflicts and to cluster those conflicts for further reflection (Schicktanz 2002). This path does not claim to provide final normative justification, but takes the reflective equilibrium as bridging between practice and theory, context and premise.

differs sometimes radically. Other common-sense principles such as trust, responsibility, or the wish for self-development might have their own ethical quality but are not satisfactorily captured within this four-principle approach.

Our purpose here is not, however, to discredit those established avenues. Each of them has its own intellectual value and practical moral purpose. They force us to argue and reflect, instead of just persisting on our spontaneous moral intuitions or moral egoism. Our aim is rather to propose a new perspective, arguing that cross-cultural comparative bioethics can be matched with all the above mentioned approaches.

1.2 The Location of Bioethical Culture

We should not open up here a discussion of the term "culture" which is of course notoriously difficult to define. A general working definition refers to culture as a complex of ideas, views, values, and social norms which materialize in symbolic presentations. To locate or identify "culture" is evidently difficult, as it is everywhere; it is self-evident, taken-for-granted, and implicit. Impossible to fully describe or understand, we therefore seek to highlight cultural elements that become more visible, according to Bhabha (1994), when there is ongoing boundary work—negotiation, conflict, or shifting margins. We use scenarios of life and death dilemmas to trigger such negotiation within our focus groups. Bioethics itself, as a discursive space, can be understood as a place of hybridity. It is a mixture of academic, public, and politicized issues. Born out of social, health and environmental movements in the late 1970s (Evans 2010; Kaminsky 2005), it has remained a mixed gathering forum even after its institutionalization. On an epistemic level, too, one can understand bioethical problems as "moral-epistemic-anthropological hybrids" (Schicktanz 2009a). The way we encounter a "bioethical" problem is already embedded within various assumptions of how to separate private versus public perspectives, social versus individual entities, rights versus virtues as normative orientations, what counts as a person or not, etc. The metaphor of "hybridity" signifies the integrative entanglement of formerly separate spheres (normativity, epistemology, anthropology) as well its potential for conflict, subversion, and displacement.

1.3 Beyond the Academic Sphere

The second direction which substantially changed the face of bioethics in this decade is the growing interest in the "non-expert." Here, a particularly important interaction between social science and bioethics took place. In the beginning, it was mainly STS (science, technology and society) scholars who argued for the involvement of lay people, patients, or stakeholder groups for a critical and ethical assessment of science (e.g., Nowotny et al. 2001). According to them, lay contributions from an epistemological as well as normative point of view will lead to a socially robust understanding and application of science and technology. Such a participatory shift also influenced—at least slightly—the study of bioethics (Schicktanz et al. 2012). The expertocratic manner of doing bioethics is increasingly criticized for being biased, exclusive or even undemocratic (Fox and Swazey 2008; Gutman and Thompson 1997, 1998). A major impact of this change was termed as the "empirical turn in bioethics" (Schicktanz 2009b; Borry et al. 2005), which again resulted in new concerns. If bioethics engages with "society," the crucial question is: How? And if we empirically detect morally relevant differences, how does this justify particular norms?

Current debates in bioethics are still under expert domination. This seems partly self-evident as bioethics is understood to be an academic discourse, so it requires experts. However, this view is problematic for two reasons: First, there exists a public bioethics as part of social (health and environment) movements, which needs to be acknowledged or even empowered. Second, even if experts dominate the discursive field, this is ethically problematic. It undermines basic democratic principles and respect for autonomy—indeed, a principle that is highlighted by modern bioethics. Social sciences and empirical-participative bioethics try to bridge or overcome some of these restrictions by bringing those perspectives into the focus of consideration. The marginalization of lay moralities does not only take place in a traditional form of power game; it is a particular form of epistemic injustice (Schicktanz 2015). Inspired by Fricker (2009), one can argue that particular voices of members of social groups (lay persons/patients) are suppressed and excluded from the public discourse because of existing social stereotypes. According to these stereotypes, patients are "just irrational," "too emotional," "too uninformed," and "too partial" to be considered as morally relevant or morally competent.

Our work starts from these two major premises: increasing our rational reflection about our own morality by a better understanding of implicit, underlying cultural scripts; as well as engaging with those who are yet excluded, to a large extent, from the academic discourse despite the relevance of their perspectives, namely those who are directly affected. Striving to give the moral perspective of being affected a place in bioethics, this book walks a tightrope between the academic requirement to remain on a reflective level, and the strong normative claims made by those who are affected concerning how issues of genetics or end-of-life should be ethically assessed.

1.4 Comparative Bioethics: Some Reasons for the Juxtaposition of Germany and Israel

Our comparative setting is quite complex: we compare two different countries, Israel and Germany. We also compare academic (expert) discourse with public moralities. Thirdly we compare different medical areas concerning difficult ethical questions of life and death, namely genetics and end-of-life questions, such as "Do I want to know whether I have a genetic predisposition for cancer?" or "When are we allowed to stop life-sustaining machines?"

Germany and Israel, both at the cutting edge of Western medical technology, generally represent opposing poles of professional culture, regulation and policies regarding biomedicine—for example, opposing policies on stem cell research, preimplantation genetic diagnosis (PGD), genetic screening, and euthanasia. The juxtaposition of these two countries is hence expected to highlight the context of national variation and pluralism in the moral assessment of biomedicine. This overarching national contrast has already been partially studied by various scholars, most notably Hashiloni-Dolev (2007), and has been related to a few major cultural factors:

(a) Religion offers metaphors and concepts that are essential to us, especially when we confront the newest borders of our medical capabilities. Judaism and Christianity, although having a common origin, offer different sources of influence in this context. This comparison evidently demands a much more detailed analysis which we will incorporate in the following chapters in the context of each medical setting. What follows is meant merely as a prelude for this comparison. Judaism arguably presents a generally more affirmative view of medicine, for example in the context of the duty to use genetic testing to avoid suffering (Brody 2002; Waldenberg 1979; Aviner 2011). There is no similar Christian view. While both religions oppose selective abortion in principle, only the Catholic Church bans pre-implantation genetic diagnosis, with some Christian thinkers encouraging the embracing of the "suffering presence" of those who are part of our community as an inherent part of human life (Hauerwas 1986). Judaism arguably embodies an alternative model to the Christian "stewardship model" that might be opposed to extreme medical intervening, for example germ-line modification, as "playing God" (Deane-Drummond 2006; Peters 2002). A dominant Jewish perspective commands us to actively follow the example of the Creator in using our human capabilities to master nature (Soloveichik 1996) and to improve upon nature (*tikkun olam*), as in "Fill the earth and conquer it" (Genesis 1:28). The Jewish view of the "pre-embryo" (Rosner 1991) has promoted a supportive stance concerning IVF, PGD and stem cell research (Zoloth 2002; Cole-Turner 2008).

In the context of end-of-life, the picture is reversed. While Orthodox Judaism forbids passive euthanasia and distinguishes between passive withholding versus active withdrawing of treatment, Christianity allows passive euthanasia in some circumstances and does not differentiate between withholding and withdrawing of treatment (Bulow 2008). The strict limit concerning the stage of the illness (last six months of life) as well as prohibiting the interruption of a continuous medical treatment (active withdrawing of treatment) which are found in the Israeli Law of the Terminally Ill but not in the German Law, are a result of the influence of Orthodox Judaism (Hurwitz et al. 2006; Barilan 2007; Teutsch and Waxman 2002). Once again we should be reminded, however, that religious restrictions represent formal statements which follow general principles, while actual decision-making is performed by lay people in concrete situations (Popovsky 2007; Doolin and Motion

2010). The shifting of perspectives between the expert ethical discourse of abstract and formal principles and the moral argumentation of lay people is quite familiar in religious life, although it had entered bioethics only recently.

(b) Yet another important influence is the interpretation of the Holocaust, especially in the context of the particular versus the universal interpretation of the lessons of the Holocaust. While there is no singular, "national" German or Israeli view of the Holocaust, scholars have pinpointed different themes or social scripts. For example, there have been many expressions in Germany of a historical responsibility to learn and respect the lessons of the Holocaust as highly relevant to current bioethical dilemmas concerning eugenics and euthanasia (Hashiloni-Dolev and Raz 2010; Erikson 2003; Krones and Richter 2004; Schroeder-Kurth 1999; Schöne-Seifert and Rippe 1991). In contrast, scholars have focused on Israeli themes in the context of interpreting the Holocaust as more particularistic, stressing the aspects of Jewish victimization and heroism (Firer 1989; Segev 1991; Raz 1994). In this reading, the lessons of the Holocaust are not seen as highly relevant to current bioethical dilemmas concerning eugenics and euthanasia. Indeed, many Israeli Holocaust survivors emphasize the singularity of the Holocaust as an event beyond comprehension and comparison, thus cautioning against drawing bioethical lessons from the Holocaust (Leichtentritt and Rettig 1999).

The lessons of the Holocaust are also relevant for the socio-cultural framing of science and medicine in the context of past experience and future prospects (Frewer and Eickhoff 2000). In Germany, expressions of a dystopian view of medical science prevail in literature, media and public discourse (Brandt 2009). Dystopian views are expressed in criticizing the technocratization and instrumentalization of medicine, as in the horrendous case of the Nazi Doctors used in slippery slope arguments. In Israel, in contrast, there has not been such a "spilling over" effect concerning the Holocaust and medicine. Rather, scholars have linked the utopian view of science and medicine in Israel to a broader cultural script where Jewish and Zionist ideologies construe uncritical views of medical technologies in the service of national survival (Prainsack and Firestine 2006; Prainsack 2006; Raz and Schicktanz 2009a, b).

(c) Another important source of differing cultural influences is the dimension of collectivism versus individualism. There is a strong social and legal emphasis on individualism and self-determination in Germany (Kemmelmeier et al. 2003), while Israel has been studied as presenting a unique blend of individualism and collectivism (Hofstede 1985), expressed for example in the value and importance of family ties and genetic kinship (Lavee and Katz 2003).

Germany and Israel also share many similarities that enable a comparison on a macro-social level. Bioethical issues are repeatedly represented in both countries' media, public debates, commissions, and political initiatives. While the German bioethics community is more established, it is also intensively emerging in Israel in recent decades. Both countries show structural similarities in health care provision which is mainly based on a publicly funded health care scheme, despite important differences in how the health care systems are organized (see Schabloski 2008). Overall, in both countries health care access is universal, in contrast to the

American health care system for example. When it comes to basic end of life care or predictive genetic tests, neither German nor Israeli patients pay for them out of their own pocket. Moreover, both countries show basic similarities with regard to legal structure. We have taken advantage, for the purpose of the study presented here, of the fact that in the last decade Germany as well as Israel implemented new laws for end of life issues and genetic testing (see Chaps. 4 and 6).

1.5 The Book's Composition

The juxtaposition of Germany and Israel, with their painful, entangled and schismatic history/ies, provides a special focus that connects social ontology, reflexive analysis, and hermeneutic positioning. We decided to work through these dimensions by presenting for the first time in detail our methodologies and our theoretical and analytical tools that were applied during our studies. Chapter 2 sets out in more detail our methodological premises and how the concept of moral grammars and cultural scripts are theoretically guiding this approach. Moreover, we will introduce three theoretical concepts that informed our analytical lenses: the notion of "being affected," consistent with valuing embodied experience and situated knowledge; the concept of "responsibility" as a meta-ethical principle which allows to identify, determine and classify moral assessments made by lay persons in social conversations; and "risk" as a necessary theoretical tool for understanding the normative and epistemological framing of modern medicine.

In Chaps. 4 and 5 we focus on the medical setting of genetic tests of adults (for late-onset diseases such as breast, prostate or colon cancer) and compare the moral grammars we found in Germany and Israel. While Chap. 4 explains the cultural and medico-legal frameworks for both countries, Chap. 5 lays out the major findings of how lay and affected people assess, value and conceptualize this new medical opportunity.

In Chaps. 6 and 7 the focus is on our second medical setting—the current debate on end-of-life decisions. We discuss advance directives as well as the debate over withdrawing or withholding life-supportive medical care ("active and passive" euthanasia), and physician-assisted suicide.

Chapter 8 is an attempt to do a cross-analysis of the two medical settings, looking for overarching patterns with regard to late-modern moral scripts of responsibility and dealing with risk in modern medicine.

Our concluding reflection is offered in Chap. 9. We want to encourage more scholars to conduct comparative bioethics that stems from transdisciplinary curiosity and not being afraid of the risk of taking the plunge. The objective of our bioethical comparison is not to decide which country or cultural context is more "ethical." Rather, this comparison allows to look at taken-for-granted issues in a new light, as well as to gain better understanding of the complex interplay between "universal" categories (e.g., the perspective of being affected) and cultural narratives, as both influence moral argumentations.

References

Aviner RS (2011) Testing a fetus for abnormalities: the responsum which hangs on hospital walls [Shut She'eilat Shlomo 2:312]. http://www.ravaviner.com/2011/02/testing-fetus-for-abnormalities.html. Accessed 30 Jul 2015

Barilan YM (2007) The new Israeli law on the care of the terminally ill. Perspect Biol Med 50(4):557–571

Beauchamp TL, Childress JF (2013) Principles of biomedical ethics, 7th edn. Oxford University Press, Oxford

Bhabha HK (1994) The location of culture. Routledge, London

Blank RH, Merrick JC (eds) (2007) End-of-life decision making: a cross-national study. MIT Press, Cambridge

Bleiklie IA, Goggin ML, Rothmayr C (eds) (2004) Comparative biomedical policy: governing assisted reproductive technologies. Routledge, London

Borry P, Schotsmans P, Dierickx K (2005) The birth of the empirical turn in bioethics. Bioethics 19(1):49–71

Brandt C (2009) Die zwei (und mehr) Kulturen des "Klons": Utopie und Fiktion im biowissenschaftlichen Diskurs der Nachkriegszeit. NTM Zeitschrift für die Geschichte der Wissenschaften, Technik und Medizin 17:243–275

Brandt RB (1961) Value and obligation: systematic readings in ethics. Harcourt, Brace & World, New York

Brody B (2002) Freedom and responsibility in genetic testing. Soc Philos Policy 19(2):343–350

Bruchhausen W (2003) Wider naiven Relativismus und vorschnellen Universalismus: Ethik in der Ethnomedizin. In: Lux T (ed) Kulturelle Dimensionen der Medizin: Ethnomedizin, Medizinethnologie, Medical Anthropology. Reimer, Berlin, pp 328–349

Bulow HH (2008) The world's major religions' points of view on end-of-life decisions in the intensive care unit. J Intensive Care Med 34:423–430

Candee D, Puka B (1984) An analytic approach to resolving problems in medical ethics. J Med Ethics 10:61–70

Cole-Turner R (ed) (2008) Design and destiny: Jewish and Christian perspectives on human germline modification. MIT Press, Cambridge

Cook JW (1999) Morality and cultural differences. Oxford University Press, New York

De Castro LD (1999) Is there an Asian bioethics? Bioethics 13:227–235

Deane-Drummond C (2006) Genetics and Christian ethics. Cambridge University Press, Cambridge

Doolin B, Motion J (2010) Christian lay understandings of preimplantation genetic diagnosis. Public Underst Sci 19(6):669–685

Erikson S (2003) Post-diagnostic abortion in Germany: reproduction gone awry, again? Soc Sci Med 56:1987–2001

Evans J (2010) Contested reproduction: genetic technologies, religion, and public debate. University of Chicago Press, Chicago

Firer R (1989) Sochnim Shel Halekach/Agents of morality. Hakibbutz Hame'uchad, Tel Aviv (in Hebrew)

Fox RC, Swazey JP (2008) Observing bioethics. Oxford University Press, New York

Frewer A, Eickhoff C (2000) "Euthanasie": Die Interaktion von Medizingeschichte und Medizinethik. In: Frewer A, Eickhoff C (eds) Euthanasie und die aktuelle Sterbehilfe-Debatte. Die historischen Hintergründe medizinischer Ethik, Campus, Frankfurt, pp 13–24

Fricker M (2009) Epistemic injustice: power and the ethics of knowledge. Oxford University Press, Oxford

Gauthier CC (2002) The virtue of moral responsibility in healthcare decisionmaking. Camb Q Healthc Ethics 11:273–281

Gutmann A, Thompson D (1997) Deliberating about bioethics. Hastings Cent Rep 27(3):38–41

Gutmann A, Thompson D (1998) Democracy and disagreement. Harvard University Press, Cambridge
Hashiloni-Dolev Y (2007) A life (un)worthy of living: reproductive genetics in Israel and Germany. Springer, Dordrecht
Hashiloni-Dolev Y, Raz A (2010) Between social hypocrisy and social responsibility: professional views of eugenics, disability and repro-genetics in Germany and Israel. New Genet Soc 29(1): 87–102
Hauerwas S (1986) Suffering presence. University of Notre Dame Press, Notre Dame
Hofstede G (1985) The interaction between national and organizational value systems. J Manage Stud 22(4):347–357
Hurwitz P, Picard J, Steinberg A (eds) (2006) Jewish ethics and the care of end-of-life patients: a collection of rabbinical, bioethical, philosophical and juristic opinions. KTAV, Jersey City
Jasanoff S (2007) Designs on nature: science and democracy in Europe and the United States. Princeton University Press, Princeton
Kaminsky C (2005) Moral für die Politik. Eine konzeptionelle Grundlegung der Angewandten Ethik, Mentis, Paderborn
Kemmelmeier M, Burnstein E, Krumov K et al (2003) Individualism, collectivism, and authoritarianism in seven societies. J Cross Cult Psychol 34(3):304–322
Krones T, Richter G (2004) Preimplantation genetic diagnosis (PGD): European perspectives and the German situation. J Med Philos 29(5):623–640
Lavee Y, Katz R (2003) The family in Israel: between tradition and modernity. Marriage Fam Rev 35:193–217
Leichtentritt RD, Rettig K (1999) Meanings and attitudes toward end-of-life preferences in Israel. Death Stud 23(4):323–358
Mackenzie C, Stoljar N (eds) (2000) Relational autonomy: feminist perspectives on autonomy, agency and the social self. Oxford University Press, Oxford
Nowotny H, Scott P, Gibbons M (2001) Re-thinking science: knowledge and the public in an age of uncertainty. Polity Press, Cambridge
Peters T (2002) Playing God?. Genetic determinism and human freedom, Routledge, London
Popovsky M (2007) Jewish perspectives on the use of preimplantation genetic diagnosis. J Law Med Ethics 35:699–711
Prainsack B (2006) Negotiating life: the regulation of embryonic stem cell research and human cloning in Israel. Soc Stud Sci 36(2):173–205
Prainsack B, Firestine O (2006) "Science for survival": biotechnology regulation in Israel. Sci Publ Policy 33(1):33–46
Raz A (1994) Rewriting the Holocaust: an Israeli case study in the sociology of the novel. In: Stone RA, Zenner W (eds) Critical essays on Israeli social issues and scholarship. Books on Israel, vol. 3. State University of New York Press, Albany, pp 9–29
Raz A, Schicktanz S (2009a) Lay perceptions of genetic testing in Germany and Israel: the interplay of national culture and individual experience. New Genet Soc 28(4):401–414
Raz A, Schicktanz S (2009b) Diversity and uniformity in genetic responsibility: moral attitudes of patients, relatives and lay people in Germany and Israel. Med Health Care Philos 12(4):433–442
Rosner F (1991) Modern medicine and Jewish ethics, 2nd edn. KTAV, Hoboken
Schabloski K (2008) Health care systems around the world. http://www.itup.org/Reports/Fresh%20Thinking/Health_Care_Systems_Around_World.pdf. Accessed 12 Jun 2015
Schicktanz S (2002) Organlieferant Tier?. Medizin- und tierethische Probleme der Xenotransplantation, Campus, Frankfurt/New York
Schicktanz S (2009a) Medizinethik-Aktuelle Trends und Neuorientierungen. In: Weidmann-Hügle T, Christen M (eds) Ethikdialog in der Wissenschaft. EMH Schweizerischer Ärzteverlag, Basel, pp 17–31
Schicktanz S (2009b) Zum Stellenwert von Betroffenheit, Öffentlichkeit und Deliberation im empirical turn der Medizinethik. Ethik in der Medizin 21(3):223–234
Schicktanz S (2015) The ethical legitimacy of patient organizations' involvement in politics and knowledge production: epistemic justice as a conceptual basis. In: Wehling P, Viehöver W,

Koenen S (eds) The public shaping of medical research: patient associations, health movements and biomedicine. Routledge, London, pp 246–264
Schicktanz S, Schweda M, Wynne B (2012) The ethics of "public understanding of ethics"—why and how bioethics expertise should include public and patients' voices. Med Health Care and Philos 15:129–139
Schöne-Seifert B, Rippe KP (1991) Silencing the singer: antibioethics in Germany. Hastings Cent Rep 21(6):20–27
Schroeder-Kurth T (1999) Screening in Germany. In: Chadwick R, Shickle D, ten Have H et al (eds) The ethics of genetic screening. Kluwer, Dordrecht, pp 81–89
Segev T (1991) The seventh million: the Israelis and the Holocaust. Hill and Wang, New York
Singer P (2003) Voluntary euthanasia: a utilitarian perspective. Bioethics 17:526–541
Soloveichik JB (1996) The lonely man of faith. Aronson, Northvale
Tao J (ed) (2002) Cross-cultural perspectives on the (im)possibility of global bioethics. Kluwer, Dordrecht
Tassicker R, Savulescu J, Skene L et al (2003) Prenatal diagnosis requests for Huntington's disease when the father is at risk and does not want to know his genetic status: clinical, legal, and ethical viewpoints. BMJ 326(7384):331–333
Teutsch D, Waxman D (2002) Behoref hayamim/In the winter of life: a values-based Jewish guide for decision making at the end of life. Reconstructionist Rabbinical College Center for Jewish Ethics, New York
Waldenberg EY (1979) Responsa Tzitz Eliezer, vol. 13. Self-published, Jerusalem, sect 102 (in Hebrew)
Zoloth L (2002) Reasonable magic and the nature of alchemy: Jewish reflections on human embryonic stem cell research. Kennedy Inst Ethics J 12:165–193

Chapter 2
Setting the Methodological Scene: The Value of Explication and Pluralization of Moral Grammars

Bioethics is currently experiencing a tremendous change as it is moving from its previous, formative, "neutral view from above" of principle-based and theory-oriented ethics to a more local, dynamic, and grounded "view from here" (De Vries et al. 2006). The thrust of this shift is derived from contrasting the previously dominant expert discourse with perspectives of non-academics, non-professionals, lay or affected persons. As a possible synthesis of these opposing views we would like to add a third view. It will be substantiated in the ensuing chapters of doing socio-empirical bioethics "through the looking glass." The act of looking at the image of our familiar culture as it is reflected on by the opposite side has the potential to produce fertile epistemological distancing, a hermeneutic repositioning from which the familiar can be seen in a new light. The metaphor of looking glasses may also be applied to the various microscopic puzzles (as well as ocular regimes of surveillance) faced by people when confronted with medical complications.

The basic methodology underlying our approach can be described as an attempt to come closer to what Habermas (1979) suggested as methodology of "rational reconstruction" of our implicit moral practice. As Habermas pointed out, our social practice of communication is characterized by argumentation—particularly in cases of disagreement. Such arguments can relate to our social world of moral conflicts, (e.g., arguing for the right of privacy), our epistemic world (e.g., arguing for the theory of evolution) or even our inner world (e.g., arguing that we are sad and disappointed). When we start to argue, we enter (according to Habermas) into a discourse. Moral discourses, in particular, force us to question our implicit assumptions or to rationalize hidden presumptions. In other words, we activate our knowledge and interpretation about relevant "cultural scripts." We borrow the term from sociologist Seale (2000). He used the term "cultural scripts" to address narratives that contain cultural representations of health and illness, for example of a good life and a good death. These cultural representations are disseminated by various means, including official discourse, the media, and so on. Seale (2000), for example, has identified four cultural scripts of dying in Anglophone countries:

A.E. Raz and S. Schicktanz, *Comparative Empirical Bioethics: Dilemmas of Genetic Testing and Euthanasia in Israel and Germany*, SpringerBriefs in Ethics, DOI 10.1007/978-3-319-32733-4_2

modern medicine, revivalism, an anti-revivalist script, and religious scripts. Even though much of dying is defined within the institutional and technical terms of reference of modern medicine, "revivalist" scripts emphasize the psychological aspects and subjective implications of death, arguing that personal acceptance of dying is an important site for the expression of individual autonomy and even personal growth. In contrast, the anti-revivalist script prefers a "closed awareness" of dying (such as not knowing one's diagnosis and/or prognosis), rejecting the more psychologically-oriented services of hospice nurses, social workers, or other professional supporters. Finally, religions all have their cultural scripts of death reflecting the religious view of (passive/active) euthanasia in the context of the sanctity of life and "not playing God" versus self-determination. Seale stresses that such scripts are not determinative, but rather should be regarded "as raw materials that are strategically (though not always consciously) used in particular situations" (1998, p. 68). People choose between different cultural scripts to deal with their particular circumstances.

2.1 Cultural Scripts Between Generalization and Enacting

While cultural scripts provide broad frames with which to discuss issues of health and illness, they should not be taken as statements of actual behavior, or confused with the ways that people think and act. One of the characteristics of post-modern societies, according to many sociologists, is the lack of master-statuses and master-narratives; the empirical study of ethics should hence examine the various kinds of scripts that are implemented. Even though religion, class or gender still may each have their corresponding "master-narrative," expressed in social stereotypes, rules, and traditions, their individual expression is often composite and versatile. When lay people express a desire for a meaningful death that is timely, pain-free, and appropriate to the way they have lived, this desire may practically interconnect various elements from different cultural scripts, including modern medicine, revivalism, and religion. Cultural scripts are therefore considered in this study as repertoires of metaphors that lay people may draw upon as they use, modify, combine, or ignore these cultural scripts.

An illustration of this complexity is provided by considering religion as a cultural script. Religious teachings are often regarded as having direct influence on decisions regarding the life course of believers, from reproduction to dying. No doubt religious faith contributes impressively to the cultural variation in attitudes towards reproduction, health, and dying. There are, however, at least three arguments that seriously undermine such generalizations. First, religious denominations such as Christianity or Judaism never imply a singular and monolithic view. For example, whereas higher religiosity in the Christian context often correlates with more skeptical/hostile attitudes towards new genetic technologies (Evans 2012), it was found that British Christians perceived religious arguments in a less constraining manner when confronted with concrete scenarios relating to their own life-worlds (Prainsack and Spector 2006). Second, religious restrictions represent formal statements which follow general principles, while actual decision-making is performed by lay people in concrete situations. As in the case of other cultural factors which are the source of general scripts, such as class, gender or nationality, religion can certainly provide a more restrictive or permissive backdrop for individual decision-making, but it does not necessarily predict it. Among Muslims in Britain, for example, as many as 67 % said that they would consider abortion following a prenatal diagnosis of thalassaemia, in contrast to the Muslim stance which bans abortion (Atkin et al. 2008).

Third, the articulation of the religious script is made, as in the case of all cultural scripts, by lay people in the context of other cultural, political and historical narratives, which shape their attitudes. In this book, we use the case of the modern-religious—Jews and Christians—to provide an actual setting for these quandaries. Our initial premise is that when shifting from the declarative to the more practical and personal, we might find a more pragmatic outlook amongst participants. It is possible that when making a decision regarding life and death, responsibilities for one's own future, as well as one's family, are seen as more influential than religion per se (Atkin et al. 2008).

Another important cultural script we work with and question at the same time is the script of "medicalization." With medicalization we refer here to Peter Conrad's term of describing a hegemonic, historical process of redefinition, power, and construction (Conrad 1975). The driving force, or biopower, is modern medicine. It is an area, both rational and pragmatic, which redefines, even in very subtle forms, all kind of issues in life as "medical issues" (Rose 2006). Already around the 1970s, "medicalization" was used by sociologists and anthropologists to denote medical, scientific understandings of human behaviors that define these behaviors as *problems* requiring surveillance and control, through treatment or management (Zola 1972; Illich 1975). Medicalization in this broad sense has occurred across many areas of human life, from "deviant" behaviors such as mental illness to "normal" life processes such as pregnancy, childbirth, and even partner choice. Genetic research and testing, an issue we focus on in this book, can be seen as an extension of the medicalization of human life by means of its clinical applications while, simultaneously, medicine is increasingly geneticized. It has for example been claimed of American breast cancer patients that the "new" genetics is medicalizing kinship by promoting inter-family discussion of genetic risk information (Finkler 2001).

The term "geneticization" was first coined to refer to the prioritizing of genetic over other understandings of human behavior (Lippman 1992, 2003). It is closely related to "genetic essentialism," a phrase used to denote scientific discourse "with the potential to establish social categories based on an essential truth about the body" (Franklin 1993, p. 34). Scholars have elaborated on the idea of the gene as a "cultural icon," suggesting that cultural representations of genetics have a life of their own, independent of the scientific research that gives rise to them (Nelkin and Lindee 1995). The thesis of geneticization has since been challenged on grounds that include the fact that empirical evidence to support it is thin: the increasing use of genetic technologies in medical practice is not in dispute, but this does not necessarily entail a widespread acceptance of a social script of deterministic or essentialist genetics (Hedgecoe 1998).

A key difficulty with medicalization and geneticization is that this terminology generally implies that lay people passively accept and use biomedical knowledge and its associated technologies, and that medical and genetic models increasingly dominate our understandings of human behavior. For us, this is something to examine rather than a premise to be taken for granted. This study asks whether acceptance and dominance of biomedical models and technologies is uniformly and unidirectionally occurring in relation to genetic risk and life plans. We expect to encounter forms of resistance and counteractions not only in the context of genetic testing but also in our second context of inquiry, end-of-life care. For example, the striving for advance directives, the discussion about respecting the patient's will to withdraw and withhold medical care, must also be seen as a form of resistance towards the modern project of medicalization.

2.2 Methodological Engagements: From Cultural Scripts to Narrative Investigation (and Back Again)

The analysis of the deliberations in the focus groups generally demonstrated that common narratives reflecting cultural scripts are created and maintained through inter-personal discourse. They are not fixed or persistent but rather dynamically and discursively (and sometimes ephemerally) negotiated as particular lay people respond to particular constraints of postindustrial technology, institutions, demographics, and notions of self. In our experience, group discussions are a very useful setting for this purpose and can be understood as an experimental form of a "mini" public discourse. Methodologically, we are not excluding other qualitative methods such as interviews or even observation. However, in these latter sources of eliciting attitudes, the researcher might have more influence on what is being said. Focus groups provide a dynamic, "natural" discourse resulting from the participants' different characters and inter-personal exchange. When interested in the hermeneutic reconstruction of arguments, hence, this qualitative methodology is particular fruitful.

According to Glaser and Strauss' (1967) classic approach to grounded theory, constant iterations of comparative analysis lead to inductive themes representing patterns of similarity and difference. Theory, or the meaning-building process of hermeneutic understanding, should not, however, be restricted to the text and its elements. The puzzle is much bigger. Meanings are deeply embedded in cultural scripts, or to use the terminology of Haraway (1988): all knowledge is situated. Along with the classic, positivistic assumption of text as covering some formal, fixed, authoritative message (to be exposed through grounded theory), we also work through these texts to expose them as knowledge situated within other, multifaceted cultural scripts. The texts produced by our focus groups are therefore treated as both "readerly" and "writerly" (Barthes 1975). As readerly texts, they should be analyzed to convey the message of the narrator. As writerly texts, they call out for multiple interpretations by us, demanding that we make an active effort in completing the text which is always a fragment of knowledge that is never morally absolute or per se valid, connecting it with broader cultural scripts.

In terms of ethical analysis, our methodology draws on the concept that the empirical social study of attitudes provides descriptive "facts" that can be understood as normative statements (Rehmann-Sutter et al. 2012; Haimes 2002). The convergence of social science and ethics intended by this approach is both of epistemological and methodological nature (Haimes 2002; Haimes and Williams 2007). This allows integrating those perspectives which are often neglected or marginalized in the dominant expert discourse (Schicktanz 2012).

First, we add to the expert discourse of abstract ethical principles and formal policies the moral arguments of lay people, which are often ambivalent, informal and "unprincipled"—a morality without theoretical foundation which is nevertheless the morality we "live by." Second, we add a methodological focus on social context—in our case lay, affected, religious, and national groups. In addition to providing empirical data for ethical analysis, the sociological analysis of these focus groups enables new questions to be asked, such as "why are these issues defined as ethical concerns by these people in these times and these places?" (Haimes 2002). Such questions can then be further discussed as indicators of broader concerns and comparative trends within Germany and Israel. The juxtaposition of the two countries is expected to highlight the context of national variation and pluralism, as well as to offer a more fine-tuned examination of group diversity and similarities within the contexts of being affected and of religiosity (Raz and Schicktanz 2009a, b). This methodological design is used to examine how cultural (national and religious) contrasts exist alongside shared positions which might reflect a common sense of being affected by disease-based experiences.

Our methodological design is used to examine how cultural (national and religious) contrasts exist *alongside* shared positions which might reflect a common sense of *being affected* by disease-based experiences. Since in both countries medical care is technically highly advanced providing intensive care and life-prolonging measures, similar contexts and decisions about treatment and health care within the medical setting can be assumed. In order to qualitatively explore these questions, we conducted focus groups with modern-religious adults, affected people, and secular (non-affected) people, in Germany and Israel. The biomedical issues discussed in each group focused on end of life and genetic testing. Similar scenarios concerning these topics were used for the semi-structured discussion. In total, 5 focus groups (N = 48) on genetic testing were conducted in a preliminary study between 2005-7.[1] Using the same methodology, 16 focus groups (N = 60) on

[1]All groups were recruited, organized and conducted in early 2005, following the guidelines for setting and content of the EU Project "Challenges of Biomedicine," Contract No. SAS6-CT-2003-510238.

end of life and genetic testing were conducted as part of the GIF-sponsored study between 2009–2012.

Focus group discussions were preferred for this study because they allow for collecting detailed material, in a purposeful and manageable way, over a short period of time (Carter and Henderson 2005; Morgan 1997). Furthermore, the scenario-based discussion which develops in a focus group combines speculative enquiry concerning normative values and assumptions in relation to specific examples in which legitimation occurs within a social context. Our focus group discussions were composed in a manner that reflected our interest in understanding personal experience with a genetic disease and with the medical system. In both countries we recruited respondents who were either patients or close relatives of a patient (i.e., "affected") or had no experience of a particular disease, test, or treatment (i.e., "non-affected", or "lay"; see Schicktanz et al. 2008). We also had a focus group discussion composed of modern-religious respondents (Christians in Germany and Jews in Israel) in each country, which provided a source of comparison concerning the bioethical arguments of the other (secular) groups as well as to the expert bioethical discourse, where religious arguments often play a major role.

We conducted eight focus group discussions (N = 60, 17 males (28 %), 43 females (72 %) with volunteers recruited in Germany and Israel between 2009–2012. In each country, two focus group discussions comprised people affected by genetic diseases. By affected we mean that the participants had clinical symptoms, or were diagnosed as carriers of a genetic disease, or had a first degree relative who has a genetic disease, and two focus group discussions with non-affected lay people—one group of secular lay people and another group of modern-religious respondents. Similar scenarios and questions were used in all focus group discussions. Focus group discussions in Israel and Germany usually included seven to nine people and lasted about two hours. There were at least to facilitators, always members of our research teams, who guided the discussion and encouraged people to comment on other people's statements.

In terms of education, groups were mixed with a slight tendency to higher level education, probably reflecting self-selection preferences. The participants' age range was 22–80 years with a heterogeneous composition in each focus group discussion. Respondents affected by genetic diseases were recruited from self-help groups and support organizations of and for people with genetic diseases. German and Israeli respondents not affected by genetic diseases were recruited by means of disseminated flyers and ads in urban public places (libraries, hospitals, pharmacies, cultural centers), and in modern-religious associations for recruiting the modern-religious participants.

The discussions were audio recorded, anonymized and transcribed, and translated into English. The transcripts from each country were analyzed thematically using the scientific software ATLAS.ti®. The analysis was compared cross-nationally in order to uncover discursive themes and categories of themes recurring within and across groups (Bloor et al. 2001). For the purpose of juxtaposing "lay moralities" and "expert discourse," emergent topics identified through inductive coding were further compared with the general categories of bioethical discourse gleaned from public policies and expert interviews. Following the cross-analysis of the transcripts, the preliminary coding was discussed by the whole team and cases of interpretive disagreement were clarified with additional codes and sub-codes added as needed.

Each focus group discussion began with expositional case vignette(s) demonstrating basic dilemmas and problems. This method was used to provide a concrete narrative that invites participants to imagine a real case, to consider what other information they would need to know to make a judgment, to consider the reasons and motives of significant others, and so on. Variants of the starting scenario were then offered to move the discussion. Standard questions asked in the focus group included a focus on moral and ethical attitudes, such as: who are the primary actors (family, individuals, professionals, the state and so on); who is seen as responsible for making the decision and why; who is seen as having the authority to decide and what are the sources of such authority (religion, morality, culture, ideology); what are the morally acceptable consequences of the decision made (feeling guilty/shame, being accused, and so on). The detailed information of the scenarios and open question presented will be described in the respective topical chapters.

We should also mention the limitations of using such qualitative methods. Qualitative studies are by nature limited in their generalizability and representativeness, as their value is in hypothesis- and theory-building (Whittemore et al. 2001). Furthermore, only a few scenarios can be discussed in depth in a single study. The selection of our discussion scenarios was inspired by controversial issues in the academic debate, such as the relevance of old age, clear-mindedness, suffering, and untreatable pain for end-of-life decisions.

References

Atkin K, Ahmed S, Hewison J et al (2008) Decision-making and antenatal screening for sickle cell and thalassaemia disorders: to what extent do faith and religious identity mediate choice? Curr Sociol 56(1):77–98

Barthes R (1975) The pleasure of the text. Hill and Wang, New York

Bloor M, Frankland J, Thomas M et al (2001) Focus groups in social research. Sage, London

Carter S, Henderson L (2005) Approaches to qualitative data collection in social science. In: Bowling A, Ebrahim S (eds) Handbook of health research methods: investigation, measurement and analysis. UK Open University Press, Maidenhead, pp 215–229

Conrad P (1975) The discovery of hyperkinesis: notes on the medicalization of deviant behavior. Soc Probl 23:12–21

De Vries R, Turner L, Orfali K et al (eds) (2006) The view from here: bioethics and the social sciences. Blackwell, Oxford

Evans J (2012) The history and future of bioethics: a sociological view. Oxford University Press, Oxford

Finkler K (2001) The kin in the gene: the medicalization of family and kinship in American society. Curr Anthropol 42:235–263

Franklin S (1993) Essentialism, which essentialism? Some implications of reproductive and genetic technosciene. In: De Cecco JP, Elia JP (eds) If you seduce a straight person, can you

make them gay?. Haworth Press, New York, Issues in biological essentialism versus social constructionism in gay and lesbian identities, pp 27–39
Glaser B, Strauss A (1967) The discovery of grounded theory. Aldine, Chicago
Habermas J (1979) Communication and the evolution of society. Beacon Press, Toronto
Haimes E (2002) What can the social sciences contribute to the study of ethics? Theoretical, empirical and substantive considerations. Bioethics 16(2):89–95
Haimes E, Williams R (2007) Sociology, ethics, and the priority of the particular: learning from a case study of genetic deliberations. Br J Sociol 58(3):457–476
Haraway D (1988) Situated knowledge: the science question in feminism and the privilege of partial perspective. Feminist Stud 14(3):575–599
Hedgecoe A (1998) Geneticization, medicalisation and polemics. Med Health Care Philos 1(3):235–243
Illich I (1975) The medicalization of life. J Med Ethics 1(2):73–77
Lippman A (1992) Led astray by genetic maps: the cartography of the human genome and health care. Soc Sci Med 35(12):1469–1476
Lippman A (2003) Eugenics and public health. Am J Public Health 93(1):11
Morgan DL (1997) Focus groups as qualitative research. Sage, Thousand Oaks
Nelkin D, Lindee S (1995) The DNA mystique: the gene as a cultural icon. Freeman, New York
Prainsack B, Spector TD (2006) Twins: a cloning experience. Soc Sci Med 63(10):2739–2752
Raz A, Schicktanz S (2009a) Lay perceptions of genetic testing in Germany and Israel: the interplay of national culture and individual experience. New Genet Soc 28(4):401–414
Raz A, Schicktanz S (2009b) Diversity and uniformity in genetic responsibility: moral attitudes of patients, relatives and lay people in Germany and Israel. Med Health Care Philos 12(4): 433–442
Rehmann-Sutter C, Porz R, Leach-Scully J (2012) How to relate the empirical to the normative: toward a phenomenologically informed hermeneutic approach to bioethics. Camb Q Healthc Ethics 21(4):436–447
Rose N (2006) The politics of life itself: biomedicine, power, and subjectivity in the twenty-first century. Princeton University Press, Princeton
Schicktanz S (2012) Epistemische Gerechtigkeit. Sozialempirie und Perspektivenpluralismus in der Angewandten Ethik. Dtsch Z Philos 60(2):1–15
Schicktanz S, Schweda M, Franzen M (2008) "In a completely different light"? The role of being affected for epistemic perspectives and moral attitudes of patients, relatives and lay people. Med Health Care Philos 11(1):57–72
Seale C (1998) Constructing death: the sociology of dying and bereavement. Cambridge University Press, Cambridge
Seale C (2000) Changing patterns of death and dying. Soc Sci Med 51:917–930
Whittemore R, Chase SK, Mandle CL (2001) Validity in qualitative research. Qual Health Res 11(4):522–537
Zola I (1972) Medicine as an institution of social control. Sociol Rev 20:487–504

Chapter 3
Applying the Theoretical Tools: Being Affected, Responsibility, and Risk

In this study we are using three analytical lenses in order to chart one territory of empirical bioethics. These are: the perspective of being affected, responsibility, and risk. It is perhaps not a coincidence that all three constructs are "experience-near," as they are built on the attitudes of lay people in everyday situations. We do not argue that these particular three constructs should be employed in each and every study of empirical bioethics. We focus on them because of their relevance to the medical settings we studied, and because we found them very fruitful for our comparative analysis. Arguably, these three constructs have external validity and relevance beyond the particular study reported here. We elaborate on this claim in the conclusion.

3.1 The Participative Turn: From the Expert to the Affected Person

One of the interesting propositions which our study explores is how moral discourses are embedded not only in the social worlds of particular groups but also in their phenomenology or lived experience. This lived experience is particularly interesting for our study as it is an individual and a social experience at the same time: as we experience pain or suffering, we feel moments of total individualism—nobody else can really understand how we suffer. But the experience of suffering also creates new social identities: People who are affected by certain medical conditions can conceive of themselves as a collective, they share expressions of this particular experience that cannot be shared by others who are non-affected. This shaping of a collective identity and "we-feeling" is usually based on a bodily, emotional and epistemic self-ascription which is an important condition for the political-social movement of patient organizations (Schicktanz et al. 2008; Schicktanz 2015). However, as in all collective identity ascriptions, this

A.E. Raz and S. Schicktanz, *Comparative Empirical Bioethics: Dilemmas of Genetic Testing and Euthanasia in Israel and Germany*, SpringerBriefs in Ethics, DOI 10.1007/978-3-319-32733-4_3

self-ascription always depends on ascriptions by others (Appiah 2005). In the past, patient groups have already shown that they can develop impressive lay expertise (Epstein 1996) which has significantly contributed to research and therapy, as well as to changes in the medico-legal discourse regarding people with disabilities.

Moral grammars and cultural patterns can arguably be changed by the lived experience of being affected, considered by some as a universalistic factor (Kleinman et al. 1997; Kleinman 1999). Hence, the perspective of "being affected" is introduced as a theoretical category. It encompasses both the epistemic and the normative dimension of a lay person's relevance in the medical-health context. The term "lay person" emphasizes the epistemic aspect of not necessarily possessing any prior scientific education and accreditation, while the complex term "being affected" implies the claim to an epistemic and normatively exclusive position, based on two factors: having experience *and* bearing the consequences (see also Schicktanz et al. 2008). The term has two basic but only partly related meanings (Rammstedt 1981): On the one hand, "being affected" refers to a causal relationship; the (mainly negative) impact an event has on a person or a group of people. It is independent of the awareness of the person(s) involved. You can be affected from a toxic dose of air pollution without knowing so. Secondly, "being affected" denotes an emotional reaction to an event. This will usually include emotions such as compassion, care or empathy. Thus someone can be affected by the pain or discrimination which ill or disabled persons experience without actually having this disability. This kind of being affected becomes explicitly relevant when we argue for solidarity and social responsibility. The two meanings of being affected are structurally different, but they are nevertheless connected: Feeling empathy with those who claim to suffer will be stronger or more enduring if the person is affected objectively. The social health movements used referenced of "being affected" for changing social awareness and politics on behalf of social solidarity with and the attaining of social power by affected persons (Schicktanz and Jordan 2013).

We assume here that being affected probably makes a cognitive difference, constituting an epistemic authority and a strong motivation for information acquisition. In a similar manner to the knowledge claims of feminist standpoint or situated knowledge theorists, critical race theorists and disability studies perspectives (Harding 2004; Johnstone 2001), affected people may warrant an "expert status" for their own situation and concerns (Badcott 2005). This also has implications for ethical norms grounded not just in the professionals' viewpoints but also in the perspectives of patients. The experiences and moral views of affected people who already have encountered genetic testing or end-of-life decision-making situations can help to understand normative priorities and moral perceptions of other patients (Abma et al. 2012).

We suggest the term "being affected" (by a medical condition) rather than "patient" as a more inclusive term encompassing direct, indirect or potential forms of being affected by a disease or a medical condition. One could speak of potential patients because of genetic disposition, as in pre-symptomatic diagnoses or predictions. In this sense, also healthy persons with a positive predictive test result for a late-onset disease can be seen as "affected" in particular contexts. In addition, relatives or spouses who are the primary care-takers of a person with a life-threatening, chronic disease are being affected by end-of-life care decision-making in a very personal manner. In the case of collective representation, it is often hard to draw the line between patients and their relatives as care-takers, even though their experience of "being affected" is not similar. Therefore, the category of "being affected" is considered as helpful for covering the social and cultural interpretations beyond medical and phenomenological understandings.

It is important to stress, however, that our use of "being affected" differs from the "affective turn" in sociology (Clough and Halley 2007). The latter mainly focuses and equates affects with emotions such as shame, love, anger, etc. According to our definition, emotions are not the major criterion. Being affected often carries emotional implications, but we consider them as additional, rather than the sole source of epistemology.

3.2 The Normative Turn: Re-Considering "Responsibility" to Analyze Moral Grammars

Responsibility was a central theme which emerged from our preliminary analysis, and later became an anchor for the comparison of lay moralities. Our focus on lay constructions of responsibility requires a conceptualization beyond legal responsibility and the question of accountability. It needs to include issues of emotional and social relations, negotiation of rights and duties, dealing with vulnerability, insecurity and uncertainty. Responsibility claims are relational, inter-connecting several moral agents and moral objects. In addition to its inductive understanding, the concept of responsibility is not all personal and idiosyncratic; it is framed by broader sociocultural and ethical scripts. Our concepts of responsibility are embedded within different cultural grammars, interconnecting formality and relationality. In English/Latin etymology, responsibility denotes an individual emphasis on self-determination: a *responsible* person is someone who is "answerable," that is, who *responds* to accusations raised in front of a court or in parliament. However, in Hebrew etymology, responsibility reflects relational support, as in standing behind someone. During their long years in the diasporas, solidarity has become a moral obligation for Jews based on the notion that "all Jews are responsible for one another" (*kol yisrael arevim ze ba-ze*, Sanhedrin 27:2, based on Leviticus 26:9–18). Such etymological differences are not merely abstract but can have very real implications, as we will see later on. However, both understandings can be captured by what we suggest as a general formula of responsibility (see below).

In philosophy, various authors have made attempts to develop a typology to analyze the spectrum of responsibility (e.g., Baier 1991; French 1991b; Ropohl 1996). Moreover, the dominant use of the term in the legal sphere has shaped much of its related discourse in other fields (see Holl et al. 2001; Bayertz 1995; Hart 1968).[1] Responsibility was (and still is) often used in the sense of "being blameworthy" or "being guilty"—which is partly problematic as we will argue later. However, the explanation itself provides a first important hint at the formal structure of what responsibility actually means: In the legal context, responsibility meant a person was "answerable", that is, had to respond to accusations raised in front of a court or in parliament (Holl et al. 2001). Our first task is to open up the meaning of responsibility so that it is not being restricted (explicitly or implicitly) to "guilt and blame." For this purpose it is important to note, that the concept of responsibility has undergone its own bioethical evolution (here summarized from: Schicktanz and Schweda 2012, pp. 137f): Since the late 1960s, "responsibility" was a leading term in applied ethics, especially concerning the emergence of new technologies. Its use and meaning(s) went manifold paths, often overlapping each other, but the increasing awareness for dealing with risk and social expectations played a major role. Responsibility is addressed here as a particular moral model. By "model" we mean a specific moral set of rules that should guide/be relevant for a particular type of actor (sometimes as collective, sometimes an individual in a particular role). Three phases were identified[2,3]:

The first phase of "responsibility" was interestingly dominated by discourses of *collective forward-oriented responsibility*, often directed towards next generations, human kind in general and nature as such. Prominent advocates were Hans Jonas or Van Rensselaer Potter who both expressed their moral concerns about technology development on the one hand and increasing ecological crisis on the other. In this context, the Asilomar Conference on recombinant DNA molecules and its potential risks of biohazard (Berg et al. 1975) is noteworthy. It has had an impact on the wider public discussion of bioethics beyond the medical-clinical context, but concerned mainly the collective (but professional) responsibility with regard to mankind and nature.

---

[1]It is noteworthy that the English translation of Aristotle's *Nicomachean Ethics* used the word "responsible." Peter French refers to Aristotle as the classical founder of our current use of responsibility, especially with regard to the question under which conditions of knowledge, choice and voluntary will one counts as a morally responsible person, or not (see extracts of Aristotle in: French 1991a, pp. 24–38).

[2]The following description differs from Turoldo and Barilan (2008) who focus mainly on versions of professional responsibility of scholars, scientists and physicians, further categorized into context-specific, antecedent, and humanized global responsibility (see also Turoldo 2010).

[3]Here we focus mainly on a European-North American perspective. It would be important to investigate in the future national/geographical differences between and within these stages. The discussion of individual responsibility, for example, started earlier in the US context, while in Europe it occurred rather after the first health care reforms in the 1990s (see Ter Meulen and Jotterand 2008). In other regions of the world such as India, an intense bioethical discussion about professional ethics has just started (Tandon 2005; Srinivasan and Loff 2006).

As Strydom (1999, pp. 74–75) points out,

> Now, from a cultural sociological or socio-cognitivist point of view, responsibility is today precisely a new master frame. It is a cultural model of social practice that has first emerged from the dynamics of the discourse generated by the new social movements, the state and business or industry in the course of antagonistically addressing the issue of risk, and then has become established beyond them at the macro-level… Since the responsibility frame is not simply a constraining cultural cognitive structure but in particular an enabling one, it admits a wide variation of practices and interpretations.

It was within that paradigm that the notion of *genetic responsibility* was coined by Lipkin and Rowley (1974) to argue in favor of reproductive and eugenic considerations (e.g., birth control or prenatal diagnosis) because of collective responsibility (Fletcher 1974).

The second stage, from the mid-1970s to the 1990s, started as intensified discussion of *professional responsibility towards individuals.* The U.S. *Belmont Report* of 1976, seen as a touchstone for informed consent procedures and the legal protection of human research participants, defined precise responsibilities for individual researchers. The success of its theoretical enlargement by Beauchamp and Childress (1979) (1st edition)[4] contributed to an understanding of professional ethos based upon balancing various ethical principles: respecting patients' autonomy, beneficence, non-maleficence, and justice. While later on enlarged for various fields in bioethics, it is important to realize that the so-called middle-range principlist approach refers in a still very subtle way mainly to the responsibility of health care professionals, because "responsibility as such" is never properly defined. Professional responsibility was not only the dominant model in the medical contexts, but also prevailed in areas such as technology assessment (e.g., Lenk 1993; Ropohl 1996; Strydom 1999). Professional responsibility was understood as a backward-oriented model of liability ("Who did wrong?"), but also as a future-oriented model of guidance ("What shall we do?") by stressing the particularly important, powerful role ascribed to professionals.

A third phase, starting in the 1990s, deals with the *interrelation between social and individual responsibility.* This on-going trend can be seen as a reaction to political reforms in which public welfare and health care systems were successively cut back, so that the tasks and burdens of individual citizen were redefined (Ter Meulen and Jotterand 2008). According to the political philosopher Iris Young, "a key term helping to propel welfare reform in the United States in the 1990s, as well as in some other reforming states", was "personal responsibility." (2011, p. 10). She criticizes that "it emphasizes that the responsibility you have is for yourself and your family…. Each must self-sufficiently bear the costs of its choices, and has no moral right to expect help from others, even if the individual or the family should suffer harm or disadvantages" (Schmitz 1998, cited in Young 2011, p. 10). Young's conclusion even goes beyond the analysis of some public health scholars. For her,

[4]It is noteworthy that even in the latest edition of this common teaching book in bioethics, the term "responsibility" or "responsible" occurs quite often, but is nowhere explicitly discussed or defined.

the emergence of personal responsibility has an "absolving function"—"to pin responsibility on one agent and to absolve others" (p. 11). According to Ter Meulen and Jotterand, personal responsibility is now more a policy to control costs because "individualization and free choice are limited by the principles of solidarity and equity" (2008, p. 195). They contrast this with the health care system in America, where the strong tradition of individualism prevails (p. 196). They also criticize that personal responsibility does not sufficiently result in free choice and autonomous decisions of the health care "consumer."

Another line of argument in favor of personal responsibility stems from theoretical critiques concerning the "perversion of autonomy" (Gaylin and Jennings 1996). Not only the rights of individuals, but also their duties should be discussed. Especially, communitarian and feminist scholars criticized the "trumping" liberal paradigm of autonomy (Callahan 1994; Held 2006; Scully 2008). For Gauthier (2002) the emerging concept of personal responsibility is seen as a virtue "that strikes a balance between the individual and the community, between rights and obligations" (p. 276). It emphasizes the "relationship with and obligation to others" (ibid.). In their deliberations about health care choices, patients should include the consequences for their families and the community, also the question of who bears the costs. Her considerations are, however, very general. In this sense, the concept of moral responsibility is reduced to an empty placeholder for "being a moral person."

The dilemma between social and individual responsibility added another perspective, especially the controversial character of knowledge and of risk assessment. An excellent example for this is Yoder's work (Yoder 2002). He summarizes recent ambiguities of individual responsibility, especially with regard to the increasing overload of information about "healthy life styles" and concludes: "If we can control our health we can be blamed for being ill" (p. 23).

3.3 Current Theoretical Challenges of "Responsibility"

Beyond the philosophical-ethical usage of the term responsibility, it is noteworthy how in social science analysis, "responsibilization" is used as a critical notion. It is often used to condemn healthcare policies imposing the internalization of individual feelings of guilt or self-restriction with regard to one's own reproductive and preventive health decisions (Lemke 2006; Rose 2006). This critical usage, however, mirrors a reductive usage of "responsibility" for only one type of actor. Furthermore, it often reduces responsibility to the so called backward or retrospective-oriented form (see the "liability" model of responsibility, below).

Such a problematic one-sided use does not justify the total elimination of "responsibility" from our moral set of thinking, as already Young has stressed. Instead, we can offset this problem by developing alternative models of responsibility. The concept of responsibility—as suggested below—provides a way out of the problem: It allows for adopting different ethical approaches (e.g., deontological vs. utilitarian vs. care ethics) and moral intuitions by applying those as considered norms.

Recent developments in medicine open up new possibilities for planning and shaping life and death on an individual as well as social level. In the context of Seale's four scripts of the good death, for example, corresponding questions of responsibility immediately arise. In the context of the script of modern medicine we can think of questions such as: What is the responsibility of doctors, and of the institutes of modern medicine, regarding euthanasia? Should doctors, who swore to save life, be made responsible to withdraw life-saving treatment, in order to respect the autonomy of the patient who made such an advance directive? And conversely: Can society responsibly promotes active euthanasia, given the economic pressures of managed care and the chronic shortage in organs for transplantation? In the context of "revivalist" scripts, questions include: Are care-takers and family members responsible for helping the patient to accept dying, or is this ultimately an individual task? In the context of anti-revivalist, we can ask: Is the patient responsible to know their diagnosis and/or prognosis? Conversely, is it irresponsible for a patient not to know their diagnosis and/or prognosis?

Lay morality research, the study of empirical ethics, explores people's moral arguments as they consider real/hypothetical scenarios. We explore what would be the individual and social responsibilities, the feared and hoped outcomes, the potential social sources of trust or mistrust. Common ethical "principles" such as autonomy, beneficence or justice do not fully or adequately address this "long-term," "open-ended," and socially-embedded normative questions. However, this does not mean, that such principles are totally arbitrary: Instead they are already embedded in complex narratives of responsibility.

3.4 The Structure of "Responsibility": A Relational Concept of Moral Agency

Responsibility, as a moral concept, has no single definition. It is linked to different contexts in moral philosophy, law, and social politics. But the underlying concept has—as we will show—a particular structure. This structure is a "metaethical" model. Metaethics is part of moral philosophy which examines the formal conception of moral language (what makes a moral claim "moral"?); it is concerned with the question how moral concepts are formalistically composed and structured or what is the epistemological basis of moral knowledge. Here, we are mainly interested in the former question: What is the "character" of responsibility that distinguishes it from other moral concepts. It is important to note that it is the structure that we are interested in; we do not defend/apply a particular normative approach (such as utilitarian, care-ethics, or Kantian approach) concerning responsibility.

A primary trait of responsibility as a moral concept is that responsibility is a *relational conception* and by this differs strongly from other ethical paradigms such as "autonomy" or "beneficence" that are commonly used in bioethics. This *general conception* of responsibility explicitly denotes a relation between several entities. At the very least, the concept of responsibility requires three relata: A subject, an object and an instance.

These three relata can be defined within the following grammatical formula: Someone (the subject, or as we will argue *the moral agent*) is responsible for someone or something (*the object*) against someone (*moral authority*). In many cases of our analysis, this basic structure is the most dominant. However, with closer inspection, it turns out that even more—five, six or even seven—relata are necessary in order to fully reconstruct and analyze our usage of the concept "responsibility" in an adequate manner (for an overview, see Werner 2002). For the analysis of "responsibility" in lay moralities, we suggest a formula that involves the following seven relata[5]:

> Someone (*moral agent*) is in a particular time frame (*time*) retrospectively/prospectively (*temporal direction*) responsible for something/someone (*object*) against someone (norm-supervising *authority*) on the basis of certain standards (*norms*) with certain consequences (*sanctions or rewards*).

In the legal context, responsibility turns out to be a *relational concept* in which a person is being held responsible by a court/authority regarding the blameworthy outcomes of his or her actions and must await a judgment which implies legal or social consequences. In other contexts, positive outcomes and rewards can occur, too. In the following, we discuss the philosophical implications of the different items/relata (see also Fig. 3.1).

A. Forward and backward temporal direction and social notion of time frame
With regard to the explanation of this formula, the *temporal direction* and distinction between retrospective/prospective is fundamental for the understanding of "responsibility." In the classical legal context (see, e.g., Bayertz 1995), responsibility was primarily discussed in a *retrospective (backward)* perspective. The question of guilt is used to ascribe a certain course of harmful past events to a particular person who can be identified as their author. This retrospective concept of responsibility is therefore also called "causal responsibility" (Bayertz 1995, p. 55; French 1991b, p. 113) or liability model (Young 2011, p. 97).

[5]We will argue that this seven-relata formula covers the most important normative aspects of the use of different models of "responsibility" in the bioethical context. In particular contexts, more relata can be useful. *The number of relata can change according to the context and the level of explicity of a particular argument. However, we do not claim that for all arguments each relata must be defined, but after testing different levels, we concluded that these seven are covering most cases.* With regard to this relational conception, the relata can be interpreted as variables: One can theoretically fill in the different relational categories with a particular, possibly infinite number of items. Theoretically, the number of possible relations in the model is vast and can be expressed by the formula (n = number of items): n(A) * n(B) * n(C) * n(D) * n(E).

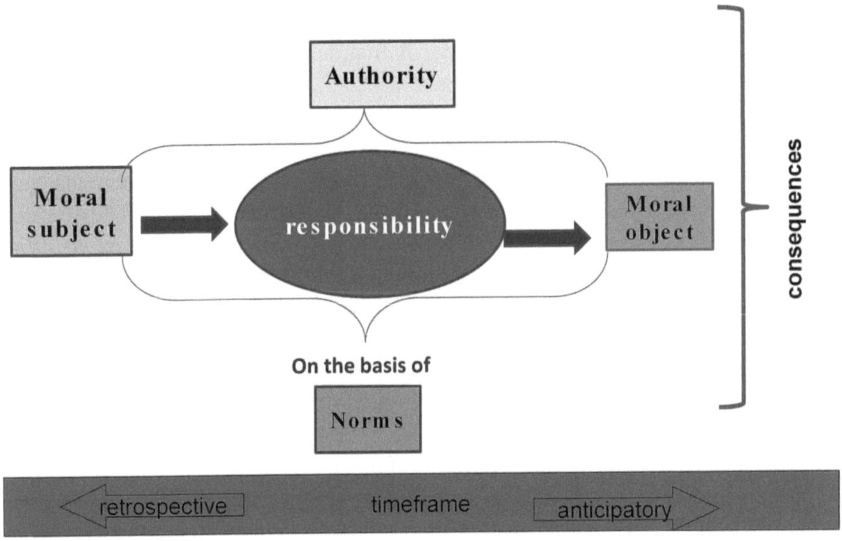

Fig. 3.1 Illustration of the 7 relata concept of responsibility

According to Bayertz, in the rise of modern society, a second, *prospective* meaning became prominent (Bayertz 1995). According to this meaning, having responsibility for an object, a person, or a group means "being in charge of" or "taking care of future events." The "actor" is hence in a particular (social, perhaps also cognitive and emotional) power position. Responsibility is associated with competence and/or power to decide (ibid.). *Prospective* (future- or forward-oriented) responsibility became important in the context of modern industrial society to deal with the complex processes of functional differentiation and network of social roles. Furthermore, technological progress made the consequences of decisions ever harder to anticipate and oversee (e.g., by the decoupling of place, time and effects, see Strydom 1999; Bayertz 1995). By this temporal shift—from backwards to forward—the whole normative meaning changed (Young 2011, pp. 76–93). Instead of judgments of blaming and guilt, prospective responsibility results in moral motivation, orientation and even justifications of socio-political engagements as a *forward-oriented* model of "taking" responsibility (p. 89).

Temporal direction is crucial for applied, context-related ethical considerations, but because of the idea of universality and equality in modern ethics, we often equate ex ante moral orientation and ex post moral judgement. For example, If person A murdered person B, A is morally condemned by C because our shared moral common sense/considered moral judgements states that it is morally wrong to murder another person. And such a judgement would be independent of the fact whether such a murder happened 20 years ago or will happen in 20 years. However, such a moral equation in time can be ex post challenged when we have to clarify the causality in a particular case: Did A intentionally drove with his car over B or did the break not work? It is also ex ante a challenge when we need to consider specifications for particular situation: Are there any morally justified exceptions? For example, what if B did try to murder A first, and A had only a chance to save his life by jumping into the car and speed up?[6]

While universal moral principles must be formulated in a general, timeless form, for the ethical analysis of a particular situation we need to differentiate, according to the time dimension: (a) when we consider retrospectively a case (similar to the court situation), we need to test for particular causality and intentions: temporal difference is built in the rational action scheme to assume that a particular action based on an actor's motives/intentions has led to a particular end (see Nowotny 1992, p. 432). (b) When we need to assess prospective cases, time defines a horizon of speculation and calculations of actions. Thus, it limits the universality of such a rule in the way we speculate and consider particular rules for exceptions or boundary conditions. However, time is not only an objective, physical category, but is also socially and culturally determined. Moreover, time perception differs subjectively. Time is therefore no absolute, but rather a multitude of perspectives and relations: the present, the past or the future do not exist as such, but are only system-relevant perspectives (Nowotny 1992, p. 436). Such a perspective helps to understand that morally relevant actions are not just actions, but events that structure our individual life course and our social life. Thus, assessing the risk of disease occurrence or cancer recurrence is often measured in survival times. However, survival time in itself has no objective meaning, but is subjectively evaluated by the individual. Assessing risk therefore depends on time dimensions in two ways. Firstly, in describing a risk, the probability of an event depends on setting a particular *time frame*: the longer or more open the time frame, the more likely that any potential harmful event can happen. Known and unknown factors that contribute to a particular event can accumulate over time. For example, it is more likely to observe an event of cancer reoccurrence the longer the time of observation. Distinguishing risks from sheer uncertainty hence needs a focus on a probability of a particular event in a clearly arranged time frame. Such a time frame is not morally neutral, but rather depends on implicit social notions of a good life (e.g., when something impacts one's life course), moral considerations of care (e.g., when something impacts one's children) or justice between generations (e.g., when something impacts the next generation).

Secondly, when assessing and evaluating a risk for individuals or a whole social group, this relies on our inventory of decision alternatives and their consequences (Hansson 2011, p. 141). Such an inventory is then evaluated by plausibility and evidence for each possibility of any event as well as a moral valuation of the outcomes. A particular step is then to consider prospective actions in their full fledge version (incl. potential effects, considering the plausibility of particular outcome in different scenarios) by adopting a so called "Future2"-position—or as Hansson

[6]What we often do is shifting this problem into an ex-post problem to clarify afterwards whether there was an "exception." However, this does not solve the problem that we accuse a person guilty or the person feels guilty until someone declares the opposite. The legal context in most liberal countries has solved this by assuming the presumption of innocence, but in many non-liberal countries, the opposite is the case.

(2011) calls it, *hypothetical retrospection*: Consider how you would assess the situation in 5, 10 or 20 years, when you then look back to the current situation and scenario 1, 2 or 3 would have happened. The different scenarios will be assessed according to our mixed judgements of underlying causality, plausibility and the moral assessment of hypothetical outcomes. Which scenario is acceptable or regrettable, inacceptable or morally wrong—where are we invariant? This moral thought experiment can be understood as a very helpful argumentative procedure. We should, however, not forget that it provides us with an illusion of an ex post judgement and therefore relies on our understanding of action-time-relations and causality. This is particularly relevant in cases when dealing with genetic risk. Here, causality and plausibility of potential outcomes (as genetic determinism) differ strongly between different diseases/phenotypes. Moreover, the time perspective impacts also differently the outcome scenarios: for example it depends whether we consider the individual life course, a member of next generation, or a whole population.

B. Individual and collective moral agents

The agent who takes responsibility or is held responsible is a moral subject. So the questions arise which agents are really moral agents, and which are not. Here, the controversy around the moral status of different subject becomes relevant: we can ask whether children, demented persons or even a company can count as moral subjects. The "agency" of such an agent should include active performance of actions as well as omissions (e.g., omitting of help for a person in a life-threatening case) (see Feinberg 1991; Held 1991).

Of course, in everyday language we often say "The virus X is responsible for the infection" or "alcohol is responsible for hepatic damage." Here are, however, the ideas of causality and responsibility confounded, instead they must be clearly separated[7]: While causality is an important part of responsibility-assignments and judgments (see below), it is problematic to equate both in the moral context. There is consensus that any moral meaning of responsibility can only be assigned to moral agents (see also Yoder 2002, p. 24). Between different moral theories exists, however, dissent about whether only individuals (persons, and if under which cognitive or moral conditions) or also collectives (e.g., families, state/nations, political groups, random groups, companies, etc.) can be classified (under which conditions) as moral agents.

The question of who counts as a moral agent is central to the need to clarify the plausibility of a particular constellation of responsibility. This debate can be traced back to Bradley (1991 [1876]). His "vulgar notion of responsibility" can be summarized in the insight, that "responsibility implies ... a capacity for acting rationally, ... to act so that your actions can be counted on" (p. 67). He states also that "a man ... is only responsible for what ... issues from an act of volition" (p. 72). For Bradley, "without personal identity responsibility is sheer nonsense" (p. 73). Harry

[7]Some of these mix-ups result from anthropomorphisms or world views in which non-human moral agents exist.

Frankfurt and John L. Mackie contributed to this debate by discussing the important role of choice and intention. Mackie's "straight rule" of responsibility ("an agent is responsible for all and only (its) intentional actions") is later modified by Peter French into responsibility based on *being willing* to do it (French 1991c, pp. 129ff). For example, a physician involved in the Tuskegee syphilis case stated that he did not have the intention to harm a patient by this experiment; his intention was to increase helpful knowledge for further vaccination research. Nevertheless, he can be held responsible for the effect of his action on the patient. It therefore seems plausible to choose *being willing* as the crucial criterion, rather than having an *intention*. This concept includes all cases where a person is willing to take some risks (e.g., harm others) and has the necessary knowledge (or could have had it).

If individuals have to *fulfill* specific conditions (such as being willing to commit an action and being aware of possible consequences) for ascribing responsibility, what does this imply for collectives? Some philosophers doubt that collectives can ever be seen as moral agents. For them, collectives must be seen as aggregations of individual agents (e.g., Ladd 1991). Hannah Arendt insisted that the liability model of responsibility should not be applied to collectives: "Where all are guilty, nobody is guilty. Guilt, unlike responsibility, always singles out; it is strictly personal" (cited in Young 2011, p. 76). Her statement has to be seen in the context of her analysis of the Eichmann lawsuit. If we accept that Eichmann and other Nazis can "excuse" their actions by referring to collective orders they had to obey, we lose all measures to ascribe liability and force individuals to take responsibility.[8]

However, whether this fear justifies a general rejection of the concept of collective responsibility is questionable. Under the impression of various ecological, medical, and economical catastrophes such as the Bhopal disaster in 1984 or the BP Deepwater Horizon oil spill in 2010, it would be implausible to ascribe responsibility for wrong doings *only* to the single worker, a captain, or a CEO, because such disaster could only happen by the interplay of various people in a system, such as a corporative or organized group. One important argument that supports the idea of collectives as moral agents points to our intuition that *sometimes* the whole (a nation, a corporation) is more than just the sum of its single parts. Those social groups, parties, a state, or a corporation can be seen as moral agents if their actions are based upon an *esprit de corps* (Cooper 1991, pp. 258–259; Schicktanz and Jordan 2013). This has to be understood as a joint commitment, and its effect cannot simply be explained by the aggregation of individual actions (French 1991d, p. 296). The argument gets even stronger where the members of organized groups[9]

---

[8]Interestingly, this position bears a serious risk of *clan liability* the other way round: where one is judged as guilty, the collective is acquitted.

[9]Held (1991) has shown that even random social groups can be seen as collectively responsible for not doing something. This requires that the members of such a random group are aware of the moral nature of a collective action (e.g., in a situation where joining to help a single woman would protect her from rape by two strong men, while a single person cannot be expected to save her). However, in health care, all relevant social actors are rather organized groups, such as professionals (physicians, nurses), politicians, or patient's advocacy groups.

have deliberatively chosen a collective action (e.g., by voting or being a voluntary member of a social group, such as a party). This group membership should be expressed as "active" commitment as Young stresses (2011, p. 137).[10] French calls it the "corporation's internal decision structure … as requisite redescription device that licenses the predication of corporate intentionality" (French 1991d, p. 298). French's definition applies not only to economical corporations but also to organized social groups, such as political parties, nongovernmental groups, and professions.

There are cases where both, the individual members and the collective, can be held responsible or only single members or only the whole. Without a model of collective responsibility the whole debate on solidary health care or the political role of professional associations, as well as such institutions as hospitals or patient organizations, cannot be addressed appropriately.

C. The moral status of moral objects

With regard to the category "objects," we assume that all moral agents can also be seen as moral objects, but it entails more entities: For example, in bioethics, we intensively discuss whether human embryos, brain-dead patients, next generations, animals, plants, or nature as a whole (for an overview: Warren 2000) must be considered as objects of responsibility according to their moral status. Such a moral status is not independent of underlying anthropological and empirical assumptions. Anthropological assumptions that are seen as morally relevant because they signify a particular status, are, for example, "experiencing pain" or "having consciousness" (e.g., Singer 1979). If we have good reasons to assume that also (some) animals feel pain or are conscious with regard to basic interests, then it is consistent to assume that they fall into the category of moral objects. However, the relatum "object" includes also sub-aspects belonging to a moral object. For example, this would entail the "healthy status" or the "body" of a person. Such entities are understood as property or essential feature of a moral object and as such are ends in themselves (Table 3.1).

D. The norm-supervising authority and instance of enforcement: Beyond the court

With regard to the category "norm-supervising *authority*," the court is a well-accepted instance. But more generally, we should think of other authorities, too, such as social peer groups, the individual conscience, or God. Such enforcement is an authority who proves and enforces the normative view. We suggest here the term "norm-supervising authority" as short version. It is the authority which decides whether a norm has been met or violated and thus the corresponding responsibility has been fulfilled or not. We note that some enforcement also can be

[10]However, there are cases where group membership is not voluntarily chosen. For example, membership in religious, gender, or ethnic groups is often ascribed from without. In this case, the application of retrospective and prospective collective responsibility is problematic.

Table 3.1 Summary of the major relata for individual, family, professional, and social responsibility

Agent (in different roles)	Object (seen as morally relevant)	(norm-proofing) Authority	Standards	Temporal direction: (negative = was not fulfilled; positive = was fulfilled)/possible consequences
Self	Self	Conscience/self	Respect for autonomy	Retrospective-negative: accept possible harm Prospective-positive: fulfilled will
			Well-being/beneficence	Retrospective-negative: guilt/blame Prospective-positive: prevention
	Family members	Conscience	Mutual respect (symmetric)/recognitions others interest	Retrospective-negative: blame/guilt/distorted relationships Prospective-positive: stabilized self-esteem/benefit of good social relations
		Social expectations	Care/love	Retrospective-negative: family distress Prospective-positive: protection of families from social-psychological burdens/increasing harmony
Family as collective/parents as couple	Individual family members (e.g., child)	Social agreement/commitment	Solidarity/care out of love (asymmetric)	Retrospective-negative: family distress Prospective-positive: social recognition/strengenetic testinghening family identity
Professional/doctor	Patient	Conscience	Respect for autonomy	Retrospective-negative: legal liability Prospective-positive: professional exclusiveness (power)/self-esteem/trust by patients
		Soft Law	Beneficience/non-harming	
		Law	Justice	
Public healthcare system	Citizens	Law	Justice and equality	Prospective-negative: waste of health care resources/inefficient allocation Prospective-positive: productivity and stability of the system
Professionals/nurses	Patient/family	Professional code/law	Respect for autonomy	Retrospective-negative: legal liability/mistrust Prospective-positive: professional power/self-esteem/trust by patients
			Care	

The temporal framework is not included because of limited layout options. It adds to each dimension a short- and long-term considerations (e.g., depends on respective disease/context)

norm-founding entities—whether we think of a court/the law or the individual conscious. This depends on pre-assumptions in different ethical theories.

Baier points to an interesting aspect: Any *instance* presupposes a task-responsibility. An instance is an authority that has to *oversee, control and judge* the responsibility of the agents. This point is important with regard to social-practical questions: If an instance has no agency power (cannot practically judge) or cannot impose sanctions, its value as a moral instance is questionable.

Thus, while the individual conscience plays a crucial role for our internal process of moral deliberation and judgment, there might be cases where referring to the psychological state of a bad conscience is insufficient. For example, the phrase *doctors are responsible (only) towards their own consciences* (as a central statement in the German professional guidelines for doctors indicates) is very unspecific. On the one hand, it points to the fundamental requirement of moral responsibility that physicians should reflect upon their own values and cannot solely rely on social norms. On the other hand—and that would be problematic—it can be interpreted in the sense that it is sufficient if someone feels shame because something went wrong.

In the case of the liability model of responsibility, we often assume that a court is *the* instance to clarify *guilt.* But on closer examination, various instances can be considered: the conscience, the profession, the family, or the broader society.

E. The plurality of moral norms and standards

The category "norms" refers to normative standards such as moral principles or legal norms (e.g., to respect others, avoid harm, save life, do justice, etc.). These principles can be very general or sometimes very concrete and context-specific (e.g., professional virtues such as confidentiality, trustworthiness, honesty but also efficiency or austerity). The standards depend, however, on their understanding as duty/obligation, right or virtue. The level of moral commitment can differ according to the standards' bindingness and acceptable excuses.

Considering the plurality of norms and standards results from a pluralistic social-philosophical image of the modern individual. While most ethicists have accepted this assumption in one way or another, conflicts with the fact that many modern approaches in ethics focus mainly on inter-individual relationships. There is a tendency to neglect the further implications of such a social and moral plurality and by extension, the political-ethical relationship between individuals, and between the individual and the state and between the individual and other social agents. Utilitarian, deontological and mid-range principle ethics share the strict liberal assumption that rights (or consequences) must be balanced (only) between individuals and reduce the state to a structure that must ensure these individual rights. Moreover, modern skepticism in ethics doubts that a decision for a particular ethical justification (e.g., utilitarian vs. deontological approach) can answer (solely) the political-philosophical problem of how, when and by whom moral problems have to be solved. Instead, procedural and content-oriented moral dimensions need to be intertwined to result into a proper moral solution. Theoretical framing need to take into account the complex social and moral web on which the different individuals in particular roles (e.g., as mother, as professional, as employee, as citizen) experience the manifold problems. This is particular true in the biomedical field in which social roles are dominating individual's decisions and self-conception's (e.g., as "patient," as "doctor," as "relative"….).

For example, the legal-ethical expert discourse brought forward the "right not to know" as well as the potentiality of psychosocial harm when discussing genetic testing of late-onset diseases, such as AD (Andorno 2004). The "right not to know," as well as its counterpart "the right to know," are both liberal expressions of protecting or ensuring the moral right of self-determination. Also this culminates in a reductionistic understanding of applied ethics into a yes/no question whether something should be morally forbidden. The dominance of such an approach in the public and international discourse originates from a legalistic-liberal orientation in bioethics. However, while this is still important, we need to ask whether such a narrow focus is ethically and intellectually satisfactory. What is left out, for example, are cases of individual or social orientation, hence questions of an individual and social good life (see Taylor 2004).

A paradigmatic case is the question of sharing genetic risk information within families. In a liberal-legalistic conception, state regulation must protect and accept private spheres, esp. the family life. It cannot stipulate particular forms of information sharing, except when there is a strong concern for a basic right (e.g., as in cases of highly infectious diseases) and hence, leave it to the individual to decide. But the moral problem of sharing and disclosure as it concerns the affected individuals is neither solved nor properly guided by such a simple permissible stance.

F. Sanctions or rewards as moral reaction to a particular outcome of an action

Finally, this category of "consequences" refers to a list of actions or judgements which are supposed to take place if a subject has or has not acted in a responsible way. Consequences are moral reactions to a particular result/end of an action. Some consequences/sanctions are external (e.g., legal or economic punishment, social exclusion or social power) and some internal (feelings of guilt or shame, having a bad conscience, or increased moral identity). Moral philosophy and applied ethics (in contrast to legal studies) have an ambiguous relationship to sanctions. While moral philosophy lengthily debates whether an action can be seen as morally wrong or right, only few discuss the consequences a moral subject should be confronted with if it did something morally wrong (e.g., Tonry 2011). The sphere of consequences is left to the social and legal sphere. However, this is not an argument for ethics to dismiss the whole concept of responsibility, but rather to be aware that we can still morally assess the actual social practice of sanctions regarding whether they are appropriate in kind and degree.

The normative complex between moral norms and sanctions is defined in terms of *proportionality*: The more binding and important an underlying moral standard is, the more demanding or powerful is the sanction that we would expect.

Furthermore, moral bindingness and consequences differ with regard to the temporal direction of an action: We can distinguish guilt, liability, or blame (or excuse) in retrospective cases, and precaution, prevention, or power and authority in prospective cases. In this sense, retro- and prospective responsibility are categorically different: Retrospective responsibility requires that something went bad or a consequence is clearly assessed as morally wrong (French 1991c, p. 135), while prospective responsibility focuses on doing morally right or at least avoiding doing wrongs.

Nevertheless, moral actions must be understood in a temporal continuum, in which backward and future-oriented views often complement each other. For example, prospective responsibility of preventive health behavior can result over time in retrospective liability models if someone claims that person X did not sufficiently prevent a disease. This idea of a negative notion of responsibility was found by Hallowell (1999), Arribas-Ayllon et al. (2008) and others in several empirical studies about genetic testing and the moral practice of disclosure or non-disclosure in affected families.

However, think about the following case:

> Lisa loves cooking, food and farmers' markets. Recently, however, she has had difficulties in remembering some of her favorite recipes. She is embarrassed that she has to look up the ingredients. Her husband has also recognized that she is getting forgetful about former conversations. Lisa is 64 and is worried that she might have Alzheimer's disease (AD). After several hours of cognitive, physiological, psychological and memory performance tests in the memory clinic, the neurologist explains that she has "mild cognitive impairment." He adds cautiously that this might be related to increasing changes in her memory and, later, to the development of AD. He suggests that it would be good to have regular follow-ups and also, if she is interested, to participate in ongoing neuroscientific research. Lisa reveals that she has already undergone genetic testing, 2 months ago. She knows that she has one copy of the APOE4 gene. She tells the doctor her own family genealogy of suspected AD by listing more than 9 family members (including her father, her grandparents, her granduncles and her great-grandparents) who all suffered from dementia. Hence, for Lisa, AD is not new but is a normal part of her family life. Her main worry is thus not her own condition but whether she might have passed it on to her two grandsons. She asks the doctor what can be done to protect the two young boys from getting this "horrible disease."[11]

The case of Lisa is not unique. The number of patients undergoing a genetic test for an Alzheimer "gene" is not so low as one might expect—given its scientific controversy and the non-existing medical option to treat Alzheimer dementia (AD). Genetic tests for Alzheimer Disease are, however, not only offered in a professional clinical setting, but most recently this interest is also served by direct to consumer (DTC) genetics companies.

---

[11]The case is adapted from field work findings of Silke Schicktanz in a California Memory clinic.

This case illustrates also very instructively how different dimensions of morality and understanding of genetics are intertwined in such everyday-life situations of a biomedical setting. The dimensions include an individual, a family, a doctor-patient, as well as a societal level.[12] However, these dimensions are hidden, rather than explicated. Her responsibility was based on a particular perception of "being at risk" for getting AD, as she interprets her family genealogy along a line of strong heredity of AD: Being at risk has motivated her to ask her local doctor to order a genetic test to test her. While she has understood that there are genes such as APOE4, her conversation with the doctor reveals also that she has perhaps not understood (or was perhaps not explained) the low level of predication associated with tests of APOE4 genes (Schicktanz and Kogel 2014). A closer reading of such a case reveals, for example, that Lisa feels morally responsible for passing genes to her grand-children. This responsibility is based on normative orientations such as caring and love, but also on epistemic concepts of risk, ageing, and hereditary, that makes the case specific; as the motivation is not about blaming or ascribing guilt.

3.5 The Epistemic Turn: The Meaning of Risk

The social, legal and ethical discourses of responsibility and risk have developed in an overlapping manner during the 20th century. 20th century's capitalism emphasized personal responsibility, and individuals rather than collective bodies were increasingly seen as taking responsibility over the personal risks of their lives, leading to the prospering of the insurance industry (Bake and Simon 2002). Our modern conceptualization of "risk" should therefore be understood as encompassing an epistemic turn. Having already become familiar with the notion of "risk" in our everyday life, most of us take it for granted. However, what we refer to as risk, ancient societies such as Rome and China called fortune, fate, or the will of the Gods. Emerging from the Italian Renaissance, along with the emergence of the Western culture of the individual, risk *defies* fate, and taking risks becomes part of the (Western) construction of the Individual (Giddens 1998; Bernstein 1998; Gregersen 2003). *Risk thinking* is a way (or a set of different ways) of bringing the future into the present, attempting to render uncertain dangers into a calculable

[12]To illustrate how such a case already entails several levels, here is a short explanation: The individual level here is Lisa's feeling of self-responsibility to look after her condition, to be tested and to inform perhaps others—without having any medical advantage as there is no proper treatment for AD. Regarding the family level, it includes the reactions of her family to such information, care about her, and her care for the two young boys (e.g., if genetic susceptibility is passed down at an increased risk, avoiding additional risk factors such as playing rugby or boxing by the boys might be wise). Professional responsibility comes into question when considering Lisa's right not know, avoiding harm by not confronting her with a "bad diagnosis" (is introducing MCI therefore a solution?) as well as in the profession's duty to intensify research to find a medical treatment. The societal responsibility covers the provision of sufficient care and avoiding social stigmatization of persons diagnosed with AD, and so on.

form. It is this form of thinking about the unpredicted as dangerous, rather than the "naturally" occurring reality, that is the heart of the modern epistemology of risk.

Risk, then, is not part of reality, but rather socially constructed. But, once risk has been constructed as an important element of our social reality, it also changes us, affects us as individuals. Ulrich Beck's writings (Beck 1992, 1999, 2000; Beck and Beck-Gernsheim 2002) chart a systematic sociological linkage between risk and modernity. Beck argues that modernity is characterized by globally manufactured or perceived risks, driven by competition, entrepreneurship, and the search for new markets. The new "risk society" where negative consequences affect all individuals (as individuals) irrespective of their social and economic circumstances, pushes us beyond reflection and towards "self-confrontation" with risk.

This study sets out to explore lay concepts of risk and responsibility in the context of bringing the future into the present and rendering uncertain dangers into a calculable form. This is taking place in the context of making plans for life (genetic testing) and death (end-of-life care). The lay negotiation of such feats provides a remarkable substantiation for theories of the "risk society" and "somatic responsibility." Genetic risks as well as end-of-life risks are great equalizers: they affect us all, basically regardless of class and gender. Health risks thus become an important element in an historical process linking modernization and individualization. The Foucauldian perspective of bio-governmentality (Rose 2006; Beck 1992; Lemke 2011) speaks persuasively about the trickling down of risk from populations (early 20ths century social hygiene) via groups (late 20th century genetic screening) and finally to individuals (21st century whole-genome sequencing). It is worthwhile to recall, once again, Yoder's (2002, p. 23) argument:

> We are told how to improve our health or reduce our risk of illness by eating properly, exercising regularly, or taking aspirin daily. While this information empowers us, it also burdens us. If we can control our health, we can be blamed for being ill.

Already 20 years earlier, Veatch (1980) pointed to this moral nexus of risk and responsibility. Health care choices and technical opportunities bear the risk that disease and disabilities are seen as someone's "fault" because individuals, parents or the state have neglected to do something against the risk. This linkage has been acknowledged, for example, in the context of smoking, exercise, low-cholesterol and low-fat diet, and other lifestyle changes that are increasingly part of our "new health morality" (Conrad 1994). With new advances in medical technology, the definition and perception of risk is extended to embrace new ways of conceptualizing uncertainty (as in the case of predictive genetic testing) or potential loss of autonomy (as in the context of end of life care and advance directives).

A study of lay moralities has the potential to go beyond previous top-down theorization of risk and responsibility in illustrating the need for further bottom-up differentiation. This can be re-connected to Beck's "reflexive modernization." A reflexively modern society is one where individuals are forced to engage as flexible producers and specialized patients-cum-consumers but must also self-critically confront, assess, and radically critique the consequences of a society where a sense of definite solutions has been lost. This self-confrontation is an inevitable part of modern life. Lay moralities are formed in this new reflexive space that has opened up. But as the chapters in this book will clearly demonstrate, reflexivity is always strongly context-bound—always embedded in cultural scripts and a particular habitus (Bourdieu 1990).

As a result of this context-embeddedness, we expect that our respondents' reflexivity—in a similar manner to reflexive modernity in general—will not produce one singular account but rather a plurality of narratives. For example, should we support a genetic test if the risk it entails can be reduced by pre-emptive treatments? The commonsense response of a reflexive member of our risk society is expected to be yes. A positive attitude toward predictive genetic testing indeed stresses the benefit of the test as leading to pre-emptive treatment. However, an equally reflexive response may be negative, emphasizing that the treatment is available irrespective of the person's knowledge about his/her genetic status, and so the test does not make a difference. Moreover, knowing too much about one's risks may be regarded as detrimental to one's health—"the more I became worried about my health, the more my health deteriorated." Should we embrace risk as part of life, test for it, confront it? Should we re-organize our life around health risks, or rather embrace these health risks as part of life? These questions carry far-reaching implications for a reflexive risk society, including the social acceptance (or not) of diversity and illness, or social regimes of health control and prevention. We continue to empirically explore these questions in the following chapters.

References

Abma T, Bruijn A, Kardol T et al (2012) Responsibilities in elderly care: Mr. Powell's narrative of duty and relations. Bioethics 26(1):22–31

Andorno R (2004) The right not to know: an autonomy based approach. J Med Ethics 30(5): 435–439

Appiah KA (2005) The ethics of identity. Princeton University Press, Princeton

Arribas-Ayllon M, Sarangi S, Clarke A (2008) Managing self-responsibility through other-oriented blame: family accounts of genetic testing. Soc Sci Med 66(7):1521–1532

Badcott D (2005) The expert patient: valid recognition or false hope? Med Health Care Philos 8:173–178

Baier K (1991) Types of responsibility. In: French PA (ed) The spectrum of responsibility. St. Martin's Press, New York, pp 117–122

Bake T, Simon J (eds) (2002) Embracing risk: the changing culture of insurance and responsibility. University of Chicago Press, Chicago

Bayertz K (1995) Eine kurze Geschichte der Herkunft der Verantwortung. In: Bayertz K (ed) Verantwortung. Prinzip oder Problem? Wissenschaftliche Buchgesellschaft, Darmstadt, pp 3–72

Beauchamp TL, Childress JF (1979) Principles of biomedical ethics, 1st edn. Oxford University Press, Oxford

Beck U (1992) Risk society: towards a new modernity. Sage, London

Beck U (1999) World risk society. Polity Press, Cambridge

Beck U (2000) Risk society revisited: theory, politics, and research programmes. In: Adam B, Beck U, van Loon J (eds) The risk society and beyond: critical issues for social theory. Sage, London, pp 211–229
Beck U, Beck-Gernsheim E (2002) Indvidualization. Sage, London
Berg P, Baltimore D, Brenner S et al (1975) Asilomar conference on recombinant DNA molecules. Science 188:991–994
Bernstein P (1998) Against the gods: the remarkable story of risk. Wiley, New York
Bourdieu P (1990) The logic of the practice. Polity Press, Cambridge
Bradley FH (1991) The vulgar notion of responsibility. In: French PA (ed) The spectrum of responsibility. St. Martin's Press, New York, pp 62–74
Callahan D (1994) Bioethics: private choice and common good. Hastings Cent Rep 24:28–31
Clough PT, Halley J (eds) (2007) The affective turn: theorizing the social. Duke University Press, Durham
Conrad P (1994) Wellness as virtue: morality and the pursuit of health. Cult Med Psychiatry 18:385–401
Cooper DE (1991) Collective responsibility. In: French PA (ed) The spectrum of responsibility. St. Martin's Press, New York, pp 255–264
Epstein S (1996) Impure science: AIDS, activism, and the politics of knowledge. University of California Press, Berkeley
Feinberg J (1991) The moral and legal responsibility of the bad Samaritan. In: French PA (ed) The spectrum of responsibility. St. Martin's Press, New York, pp 148–169
Fletcher J (1974) Genetics, choice and society. In: Lipkin MJ, Rowley PT (eds) Genetic responsibility: on choosing our children's genes. Plenum Press, New York, pp 93–100
French PA (ed) (1991a) The spectrum of responsibility. St. Martin's Press, New York
French PA (1991b) Types and principles of responsibility. In: French PA (ed) The spectrum of responsibility. St. Martin's Press, New York, pp 113–116
French PA (1991c) Fishing the red herrings out of the sea of moral responsibility. In: French PA (ed) The spectrum of responsibility. St. Martin's Press, New York, pp 129–144
French PA (1991d) The corporation as a moral person. In: French PA (ed) The spectrum of responsibility. St. Martin's Press, New York, pp 290–304
Gauthier CC (2002) The virtue of moral responsibility in healthcare decisionmaking. Camb Q Healthc Ethics 11:273–281
Gaylin W, Jennings B (1996) The perversion of autonomy. Free Press, New York
Giddens A (1998) Conversations with Anthony Giddens: making sense of modernity. Stanford University Press, Stanford
Gregersen NH (2003) Risk and religion: toward a theology of risk taking. Zygon 38(2):355–376
Hallowell N (1999) Doing the right thing: genetic risk and responsibility. Sociol Health Ill 21(5): 597–621
Hansson SO (2011) Coping with the unpredictable effects of future technologies. Philos Technol 24:137–149
Harding S (ed) (2004) The feminist standpoint theory reader: intellectual and political controversies. Routledge, New York
Hart HLA (1968) Punishment and responsibility: essays in the philosophy of law. Clarendon Press, Oxford
Held V (1991) Can a random collection of individuals be morally responsible? In: French PA (ed) The spectrum of responsibility. St. Martin's Press, New York, pp 265–274
Held V (2006) The ethics of care: personal, political, and global. Oxford University Press, Oxford
Holl J, Lenk H, Maring M (2001) Verantwortung. In: Ritter J, Gründer K (eds) Historisches Wörterbuch der Philosophie, vol 11. Schwabe, Basel, pp 566–575
Johnstone D (2001) An introduction to disability studies, 2nd edn. Fulton, London
Kleinman A (1999) Moral experience and ethical reflection: can ethnography reconcile them? A quandary for "the new bioethics". Daedalus 128(4):69–97
Kleinman A, Das V, Lock M (eds) (1997) Social suffering. University of California Press, Berkeley

Ladd J (1991) Corporativism. In: French PA (ed) The spectrum of responsibility. St. Martin's Press, New York, pp 305–312
Lemke T (2006) Genetic responsibility and neo-liberal governmentality: medical diagnosis as moral technology. In: Beaulieu A, Gabbard D (eds) Michel Foucault and power today: international multidisciplinary studies in the history of the present. Lexington, Lanham, pp 83–91
Lemke T (2011) Critique and experience in Foucault. Theory Cult Soc 28(4):26–48
Lenk H (1993) Über Verantwortungsbegriffe und das Verantwortungsproblem in der Technik. In: Lenk H, Ropohl G (eds) Technik und Ethik. Reclam, Stuttgart, pp 112–148
Lipkin M Jr, Rowley PT (eds) (1974) Genetic responsibility: on choosing our children's genes. Plenum Press, New York
Nowotny H (1992) Time and social theory: towards a social theory of time. Time Soc 1(3): 421–454
Rammstedt O (1981) Betroffenheit—was heißt das? Politische Vierteljahreszeitschrift 22 (Sonderheft 12):452–463
Ropohl G (1996) Ethik und Technikbewertung. Suhrkamp, Frankfurt
Rose N (2006) The politics of life itself: biomedicine, power, and subjectivity in the twenty-first century. Princeton University Press, Princeton
Schicktanz S (2015) The ethical legitimacy of patient organizations' involvement in politics and knowledge production: epistemic justice as a conceptual basis. In: Wehling P, Viehöver W, Koenen S (eds) The public shaping of medical research: patient associations, health movements and biomedicine. Routledge, London, pp 246–264
Schicktanz S, Jordan I (2013) Kollektive Patientenautonomie: Theorie und Praxis eines neuen bioethischen Konzepts. In: Simon A, Wiesemann C (eds) Patientenautonomie. Mentis, Münster, pp 287–302
Schicktanz S, Kogel F (2014) Genetic responsibility revisited: moral and cultural implications of genetic prediction of Alzheimer's disease. In: Prainsack B, Schicktanz S, Werner-Felmeyer G (eds) Genetics as social practice. Ashgate, Farnham, pp 199–218
Schicktanz S, Schweda M (2012) The diversity of responsibility: the value of explication and pluralization. Medicine Studies 3:131–145
Schicktanz S, Schweda M, Franzen M (2008) "In a completely different light"? The role of being affected for epistemic perspectives and moral attitudes of patients, relatives and lay people. Med Health Care Philos 11(1):57–72
Schmitz D (1998) Taking responsibility. In: Schmitz D, Goodin RE (eds) Social welfare and individual responsibility. Cambridge University Press, Cambridge, pp 1–96
Scully JL (2008) Disability bioethics: moral bodies, moral differences. Rowman & Littlefield, Lanham
Singer P (1979) Practical ethics. Cambridge University Press, Cambridge
Srinivasan S, Loff B (2006) Medical research in India. Lancet 367(9527):1962–1964
Strydom P (1999) The challenge of responsibility for sociology. Curr Sociol 47(3):65–82
Tandon PN (2005) Bioethics: an emerging discipline. Indian J Med Res 121(1):1–4
Taylor S (2004) Predictive genetic test decisions for Huntington's disease: context, appraisal and new moral imperatives. Soc Sci Med 58(1):137–149
Ter Meulen R, Jotterand F (2008) Individual responsibility and solidarity in European health care: further down the road to a two-tier system of health care. J Med Philos 33(3):191–197
Tonry M (ed) (2011) Why punish? How much?. Oxford University Press, Oxford, A reader on punishment
Turoldo F (2010) Ethics of responsibility in a multicultural context. Perspect Biol Med 53(2): 174–185
Turoldo F, Barilan MY (2008) The concept of responsibility: three stages in its evolution within bioethics. Camb Q Healthc Ethics 17(1):114–123
Veatch RM (1980) Voluntary risks to health: the ethical issues. JAMA 243(1):50–55
Warren MA (2000) Moral status: obligations to persons and other living things. Oxford University Press, Oxford

Werner MH (2002) Verantwortung. In: Düwell M, Hübenthal C, Werner MH (eds) Handbuch Ethik. Metzler, Stuttgart, pp 521–527

Yoder SD (2002) Individual responsibility for health: decision, not discovery. Hastings Cent Rep 32:22–31

Young IM (2011) Responsibility for justice. Oxford University Press, Oxford

Chapter 4
Contextualizing the Cultural and Medico-Legal Debate on Adult Genetic Testing

Genetic testing brings together, in a very concrete manner, the notions of being affected, responsibility, and risk. Genetic mutations imply risk for those who carry them, are affected by them, and may feel personal responsibility (or social pressure) to test for them. As we are all fellow mutants, each and every one of us a carrier of various mutations, we are all subjected to a growing discourse of genetic responsibilization.

4.1 The Script of "Genetic Responsibility"

Critical sociologists have already put forward the concept of "genetic responsibility" as a new emerging social norm in the field of genetic medicine. It has been highlighted in recent studies on attitudes towards predictive genetic testing of adults in the context of breast cancer and Huntington's disease (Hallowell 1999; Konrad 2003, 2005; Taylor 2004; Novas and Rose 2000; Arribas-Ayllon et al. 2008). For example, self-responsibility in genetic testing could reflect the conflict between the right not to know vs. the duty to know in cases where therapy does not exist, but a special lifestyle may help to reduce disease outbreak. This could also be linked in several ways to relational responsibility—expecting to be told about genetic test results by relatives and feeling an obligation to tell them (Hallowell 1999). In the social sciences context, "responsibility" (as self- or parental-responsibility) is often criticized as a normative regime of pastoral control. A comprehensive example of how to apply the Foucauldian perspective to the new genetics is Novas and Rose's (2000) and more recently Rose's (2006) work on genetic risk and the somatic individual. Genetic risk, according to this analysis, generates new forms of genetic responsibility, locating affected individuals within new communities of obligation and leading to the development of novel "life strategies." Moreover, "genetic responsibility" often does not differentiate between the individual's responsibility towards himself and the one towards his family which is highly problematic.

A.E. Raz and S. Schicktanz, *Comparative Empirical Bioethics: Dilemmas of Genetic Testing and Euthanasia in Israel and Germany*, SpringerBriefs in Ethics, DOI 10.1007/978-3-319-32733-4_4

As Arribas-Ayllon et al. (2011) show in their interview study, the majority of individuals as family members practiced nondisclosure: "The prospect of disclosure was expressed as a practical moral burden of calculating who to tell and when to tell them…." (p. 11). The interviewed lay persons justified nondisclosure in terms of their own guilt, the possibility of incurring blame or by pre-empting a relative's right not to know. One of the most common explanations for non-disclosure between generations was the strong desire to protect the future autonomy of children and young adults. Therefore, the authors argue "for a different sense of genetic responsibility, which is neither reducible to transcendental communication nor that of rational autonomy" (p. 20). They instead suggest "that people resist the onerous obligation of managing genetic risk and disclosing bad news to kin to escape the moral/discursive confinement of autonomy" (p. 19). They interpret this as an intensity of affect (ibid.), a pre-discursive strategy of these family members practicing non-disclosure. However, we should take much more serious what their informants gave as reasons instead of providing just academic interpretations. Here, norms such as respecting the *right not to know,* and *care for family members* to avoid socio-psychological burdens for others and the family as whole were mentioned. According to our analysis, those who disclose genetic knowledge to family members do not subsequently act upon autonomy as the leading norm; rather, they act on the norm that others have a *right to know* or in the case of non-disclosure a *right not to know.*[1] From here, different kinds of consequences can result: Blame as well as feelings of guilt are feared by family members, in both liability models.

As early as 1974, the libertarian philosopher Robert Nozick suggested that prospective parents should be free to choose the genetic features of their children. This was then and still today mainly a science fiction phantasy, as the technical opportunities are still not available to choose genetic features, but only to select between embryos. In an interesting coincidence, in the same year a book titled *Genetic Responsibility: On Choosing our Children's Genes* (Lipkin and Rowley 1974) was also published. Indeed, procreative autonomy regarding the choice of your child's genetic features may quickly turn into a normative discourse of parental responsibility. From a Foucauldian perspective, the widespread notion of genetic risk can be seen as producing self-government of individual behaviors and life strategies, which is expressed in terms of genetic responsibility (Lemke 2002, 2005).

Genetic responsibility as parental duty is not only promoted by some geneticists, but also by a few bioethicists. They argue for a moral obligation to use genetic testing for severe diseases as a minimal parental duty, and even claim that screening for the carrier-status component of that obligation might appropriately be compelled by the state through a well-designed program (Brody 2002). As Angus Clarke (1997, p. 88) compellingly argues:

[1]Within the ethics of genetics, these norms are intensively discussed and are already part of legal-political regulations (Chadwick 2009; e.g., German Law on Genetic Testing, UNESCO Bioethics guidelines).

> The use of the word "prevention" in many health care contexts will lead on to the notion of "responsibility" and thence to the stirring of the flames of blame and guilt. In the context of genetics, the corresponding notion is "reproductive responsibility"; the failure to comply with the recommended screening test is already commonly regarded as blameworthy. It would be completely unacceptable for a health service deliberately to provoke such emotions in the parents of children with serious diseases. It is simply not appropriate to approach prevention in the context of genetic diseases in the same way as the health education and the anti-tobacco campaigners lobby for the prevention of lung cancer.

But, for most of us without a clear medical family history, *ignoring* genetic information about our hereditary properties can be ethically advisable if we are willing to unconditionally nurture a child despite her genetic characteristics (Vehmas 2001).

4.2 The Practice of Adult Genetic Testing in Germany and Israel

Germany and Israel generally represent contrasting legal regulations and professional outlooks, particularly in relation to reproductive, prenatal genetic testing and pre-implantation genetic diagnosis (PGD). German genetic counsellors are more cautious regarding the use of prenatal diagnosis for selective abortion, while Israeli geneticists are often in favor of it (Hashiloni-Dolev 2007; Wertz and Fletcher 1989; Gottweis and Prainsack 2006). While German patient advocacy groups like *Aktion Mensch* are critical of prenatal testing (Hashiloni-Dolev and Raz 2010), many disability activists in Israel support it (Raz 2004). Although the German legal situation has changed in 2011 by allowing, for the first time, pre-implantation genetic diagnosis (PGD) for serious early-onset diseases, its regulation is still very restrictive in relation to Israel, where PGD is allowed for HLA typing (the creation of "sibling donors") and family balancing under specific conditions (Grazi et al. 2008).

In this study we focus on genetic testing of adults, which includes carrier and predisposition testing. This is an important area of new genetics usually seen as increasing the range of choices open to the healthcare consumer (Vallance and Ford 2003; Slowther 2008). While predictive and carrier testing are both practiced in Germany and Israel on a personal choice basis, it is only in Israel that carrier screening of adults is regulated by the state as part of a national screening program as well as community-based screening programs for high-risk ethnic groups (Zlotogora et al. 2009). It is interesting to observe how the national practice of adult carrier screening reflects cultural differences: The German model is more individualistic, driven by self-determination, whereas the Israeli model encompasses both individual carrier testing as well as collectivistic and communal, state-implemented screening targeting specific at-risk ethnic groups.

While the use of ethnicity in genetic databases and screening programs may be subject to debate, in Israel this practice has been well accepted (or at least implemented without public debate). Population screening for Tay Sachs disease among

Ashkenazi Jews was the first community oriented genetic screening of adult carriers in Israel. Later, probably as a direct result of the overwhelming success of the Tay Sachs program, Israel was the first country in which population screening was introduced for cystic fibrosis, fragile X, and spinal muscular atrophy, free of charge. Thereafter, the community of origin has been routinely used in medical genetics either for Jews, Arabs, Bedouins or Druzes living in Israel without generating any debate either within the medical community or the general public (Zlotogora 2015).

Adult carrier screening and testing in ethnic communities in Israel is of particular importance due to the high frequency of consanguinity. Similarly to the Arab population in the Middle East, consanguineous marriages are frequent among the Israeli Arabs, Bedouins and the Druze with a preference for first cousin marriages. In more than 25 % of Muslim Arabs and Druze marriages the spouses are first cousins with an additional 20 % related in other ways (Vardi-Saliternik et al. 2002). Many autosomal recessive diseases have been diagnosed among the Israeli Arabs, Bedouins and Druze, some of these diseases being particularly prevalent in a large extended family, a village, a region, or even in some cases in a whole community (Zlotogora 2010).

Perhaps the most widely-known example of successful community-based pre-marital, adult carrier screening is *Dor Yesharim*, a program developed by a Rabbi and implemented by the ultra-orthodox Jewish community in the U.S., Europe and Israel (Eckstein and Katzenstein 2001). Jewish Orthodox law (*halacha*) forbids abortion (unless the woman's life is endangered) thus making prenatal genetic testing irrelevant. Many of the marriages in this population are arranged. First and second degree consanguinity rate is relatively low (3–7 %), but there is a high rate (~25 %) of intra-community marriages, namely endogamy (Cohen et al. 2004). Dor Yesharim extends the traditional norm of pre-arranged endogamous match-making to include carrier matching for genetic diseases that are prevalent in the Ashkenazi Jewish population, such as Tay-Sachs (Raz and Vizner 2008). The system works in the following manner: Dor Yesharim representatives visit ultra-orthodox high schools and draw blood samples from students, who are then issued a code. Individuals never receive their carrier status, to avoid stigmatization and anxiety. The blood samples are screened for several genetic diseases and the results stored in Dor Yesharim's offices. When young men and women reach a marriageable age, and receive a recommendation from a matchmaker about a potential mate, they or their parents are supposed to make a phone call to a representative of Dor Yesharim, who retrieves the assigned codes for each member of the potential couple and checks to see if they are carriers of the same genetic disease. If they are, they are told that a union is *not advisable*. The only result that the tested individuals receive is either "advisable" or "nonadvisable" for marriage. In this way, most carriers would never find out what gene they carry and thereby are supposed to avoid being anxious about that information as well as being seen as defective. The match is considered to be compatible if both parties are not carriers of the same recessive trait. Each member of the couple may be a carrier for a different disorder, but this information is not revealed, as it does not affect their compatibility as a couple.

Since its inception in the 1980s, over 220,000 individuals have been tested, over 500 incompatible couples identified, and virtually no afflicted children were born (Eckstein and Katzenstein 2001) in the ultra-orthodox community in Israel, which comprises about 800,000 people (out of 5.4 million Jews living in Israel, a country of about 8 million). Recently, Dor Yesharim has aimed at increasing its activities, reaching out to other religious communities within Jewish society, such as the "modern religious" community (Frumkin et al. 2011).

In the 1990s, geneticists and epidemiologists attempted to implement a screening program to identify adult carriers in the Bedouin community in the south of Israel. This endeavor was inspired by the success of Dor Yesharim (Raz 2005). The aim of this pilot program was to allow for the use of information on the carriage of genetic diseases (starting with deafness, which was frequent in the tribe) in decisions about marriage. This intervention was not successful, one of the main reasons being that it overlooked the social importance of consanguineous marriage for community members (Raz 2005). In the last decade, efforts have been made to develop and offer carrier screening in a manner that does not interfere with the choices of individuals, but delivers genetic counselling which is targeted to the needs and risks of Muslim Arabs and Druze communities in Israel. This "community genetics" approach hinges on offering genetic testing services in community-outreach clinics, based on local genetic databases of the community, and without preaching against consanguinity per se (Zlotogora 2015).

In Germany, there are no community-based carrier screening programs that target ethnic groups, although there might be minorities where consanguinity is more prevalent. Indeed, the issue of carrier screening is discussed separately from and with much more caution than individual genetic testing (Schroeder-Kurth 1999; Enquete Commission 2002; German Ethics Council 2013). The German sensitivity regarding ethnic carrier screening has been related to the horrendous lessons of Nazi eugenics and the Holocaust. The German model is therefore restricted to individual carrier testing for adults, driven by self-determination, whereas the Israeli model encompasses both individual carrier testing as well as State-implemented community-based screening targeting specific at-risk ethnic groups. It is noteworthy, that the new concept of preconceptional carrier testing for several, even up to more than 500 rare genetic disorders, as it is provided by some internet-based firms in North America and also promoted now in the Netherlands, is rarely discussed in Germany (Inthorn et al. 2014).

In Israel, genetic screening is considered a national priority and a moral duty, legitimating the targeting of at-risk ethnic groups by government authorities and medical geneticists. In Germany (in a similar manner to other European countries and the U.S.), where genetic screening is not considered a national priority but rather an issue of individual choice, government authorities, geneticists and support groups for people with genetic diseases refrain from making recommendations to people with the genetic disease to pursue or not pursue any type of genetic testing, in line with the ethos of non-directive counseling. German support organizations recommend that genetic counseling must include information about how to live with a child with the genetic condition, and how to support the child's special needs. For the majority of Israeli organizations, this issue is still under-developed. This national difference demonstrates how the moral meaning of scientific technologies is shaped within cultural contexts. The question of whether and how to implement and use a medical technology—such as adult carrier screening and testing—is also contingent on our personal as well as communal narratives of risk and responsibility. Nevertheless, there are also differences that emanate from the medical aspects of the genetic disease in question, as the following section explicates.

4.3 Two Paradigmatic Cases of Predictive Genetic Testing for Late-Onset Diseases: Huntington's Disease and Colon Cancer

Huntington's disease is a dominantly inherited disease for which predictive testing can inform whether, but not precisely when, the disorder will manifest itself in adulthood. Testing for Huntington's disease is not generally recommended, in part because no preventive or curative treatments are available, and because Huntington's disease has a relatively low prevalence of five cases in 100,000 inhabitants (Hawkins et al. 2011). In contrast, colon cancer is a common disease in Western countries, with over 25 cases per 100,000 inhabitants. It is the second common cancer in women and third common cancer in men in industrialized countries. The rate of colon cancer with a genetic basis is estimated between 5–25 %, with overweight and malnutrition as contributing external risk factors (Duncan et al. 2008). While Huntington's disease testing implies high predictability without treatment, colon cancer testing is characterized by increased susceptibility with often successful treatment, including radio- and chemotherapy and surgery or preventive care by healthy nutrition, exercise and early detection by regular checks. In Huntington's disease, knowing one's genetic status and predicted age of onset can eliminate doubt and assist in making life plans, but the prospect of developing a fatal disease can be far more stressful than the uncertainty (Erez et al. 2010).

This may explain why only a relatively low percentage of those with a family history of Huntington's disease have opted to be tested—with a conflict whether or not to know and to tell kin (Konrad 2003, 2005; Taylor 2004). Conversely in colon cancer, those with a family history of the condition are often referred to genetic counselling (Duncan et al. 2008). Testing relatives at risk for colon cancer is recommended in Israel, particularly for preventive considerations (Rosner et al. 2009). Similarly in Germany, there is high approval of testing for colon cancer with at-risk persons holding a more favorable view of the testing (Berth et al. 2002).

For example, the different regulation of pre-implantation genetic diagnosis (PGD) in Germany and Israel arguably relates to two different paradigms of responsibility: toward the protection of embryos (in the case of German regulation illustrated by the Embryo Protection Law) and toward the perceived *promotion* of procreative autonomy. Medical technology enables in-vitro-fertilization and genetic diagnosis to produce a child with human leukocyte antigen properties matching those of an ill sibling and who is therefore a suitable stem cell donor for that sick child. Pre-implantation genetic diagnosis to create a matching donor sibling is banned in Germany, but endorsed in Israel, due to different ideas regarding the ideal relationship between the generations, and the mutual obligations between different family members (Hashiloni-Dolev and Shkedi 2007, 2010). Arguably, the German banning of pre-implantation genetic diagnosis to create a sibling donor stems from a view of family members as first and foremost individuals holding autonomous rights, with a view of the family as a potentially highly exploitive social group and hence the social responsibility to protect the rights of the yet unborn child—the pre-embryo. Until 2010, the total prohibition of pre-implantation genetic diagnosis in Germany had been inferred from the Embryo Protection Law (1990). The German law thus permits sibling "donations" by already existing siblings, but prohibits the use of pre-implantation genetic diagnosis to select for such donor siblings. The objection to such genetic screening of prospective children may also stem from a contention that it leads to a eugenic slippery slope of "designing" children. The concern is that it will result in instrumentalizing the life of these "designer babies," with possible adverse impacts on the physical and emotional welfare of sibling donors, as well as on their autonomy and dignity, or irreversibly changing the relationship between the "produced" child and the parents (Habermas 2003; Sandel 2007). Such ethical assessments call for greater measures of social responsibility in protecting the diversity of lives. In contrast, support of pre-implantation genetic diagnosis in Israel arguably stems from a view of the family as a unified body of members with similar interests and mutual responsibilities. "Consequently, pre-implantation genetic diagnosis for sibling donors is viewed in Israel as a blessing which serves all family members, rather than as a medical procedure creating family conflicts" (Hashiloni-Dolev and Shkedi 2007, p. 2090).

Arguably, no single narrative of responsibility can explain *all* the differences. The recent changes in the German regulation of pre-implantation genetic diagnosis also supports this argument, demonstrating that medico-legal regulation can be flexible, rather than being fixated upon deep-seated cultural assumptions. When we compare the national, declarative policy to the moral attitudes of lay people, additional differences evidently present themselves. In Germany, 60 percent of infertile couples supported legalizing pre-implantation genetic diagnosis for HLA-matching, but only a minority approved pre-implantation genetic diagnosis to test for non-health-related traits (Borkenhagen et al. 2007). A dialectical focus on the cultural grammars behind individual and interpersonal concepts of responsibility can provide an intriguing interface for bridging some of the gaps between experts' formal ethics of principles and our lay moralities, between ground and surface rules of moral action, and between theoretical and empirical bioethical analysis.

References

Arribas-Ayllon M, Featherstone K, Atkinson P (2011) The practical ethics of genetic responsibility: non-disclosure and the autonomy of affect. Soc Theor Health 9:3–23
Arribas-Ayllon M, Sarangi S, Clarke A (2008) Managing self-responsibility through other-oriented blame: family accounts of genetic testing. Soc Sci Med 66(7):1521–1532
Berth H, Balck F, Dinkel A (2002) Attitudes towards genetic testing in patients with HNPCC/FAP in the German population. Genetic Test 6:273–280
Borkenhagen A, Brähler E, Wisch S et al (2007) Attitudes of German infertile couples towards preimplantation genetic diagnosis for different uses: a comparison to international studies. Hum Reprod 22(7):2051–2057
Brody B (2002) Freedom and responsibility in genetic testing. Soc Philos Policy 19(2):343–350
Chadwick R (2009) The right to know and the right not to know: ten years on. In: Rehmann-Sutter C, Müller H (eds) Disclosure dilemmas: ethics of genetic prognosis after the "right to know/not to know" debate. Ashgate, Farnham, pp 9–18
Clarke AJ (1997) Population screening for genetic carrier status. In: Harper PS, Clarke AJ (eds) Genetics, society and clinical practice. Bios Sci, Oxford, pp 77–92
Cohen T, Vardi-Saliternik R, Friedlander Y (2004) Consanguinity, intracommunity and intercommunity marriages in a population sample of Israeli Jews. Ann Hum Biol 31(1):38–48
Duncan RE, Gillam L, Savulescu J et al (2008) "You're one of us now": Young people describe their experiences of predictive genetic testing for Huntington disease (HD) and familial adenomatous polyposis (FAP). Am J Med Genet Part C 148C:47–55
Eckstein J, Katzenstein H (2001) The Dor Yeshorim story: community-based carrier screening for Tay-Sachs disease. Adv Genet 44:297–310
Enquete Commission on Law and Ethics of Modern Medicine (2002) Final report. German Parliament, Berlin (in German)
Erez A, Plunkett K, Sutton VR et al (2010) The right to ignore genetic status of late onset genetic disease in the genomic era; prenatal testing for Huntington disease as a paradigm. Am J Med Genet Part A 152A:1774–1780
Frumkin A, Raz A, Plesser-Duvdevani M et al (2011) "The most important test you'll ever take?" Attitudes towards confidential carrier matching and open individual testing among modern-religious Jews. Soc Sci Med 73(12):1741–1747
German Ethics Council (2013) Future of genetic diagnosis—from research to clinical practice. http://www.ethikrat.org/files/opinion-the-future-of-genetic-diagnosis.pdf. Accessed 03 Aug 2015
Gottweis H, Prainsack B (2006) Emotion in political discourse: contrasting approaches to stem cell governance in the US, UK, Israel and Germany. Regen Med 1:823–829
Grazi R, Wolowelsky JB, Krieger DJ (2008) Sex selection by preimplantation genetic diagnosis for nonmedical reasons in contemporary Israeli regulations. Camb Q Healthc Ethics 17:293–299
Habermas J (2003) The future of human nature. Polity Press, Cambridge
Hallowell N (1999) Doing the right thing: genetic risk and responsibility. Sociol Health Ill 21(5): 597–621
Hashiloni-Dolev Y (2007) A life (un)worthy of living: reproductive genetics in Israel and Germany. Springer, Dordrecht
Hashiloni-Dolev Y, Raz A (2010) Between social hypocrisy and social responsibility: professional views of eugenics, disability and repro-genetics in Germany and Israel. New Genet Soc 29(1): 87–102
Hashiloni-Dolev Y, Shkedi S (2007) New reproductive technologies and family ethics: pre-implantation genetic diagnosis for sibling donor in Israel and Germany. Soc Sci Med 65(10): 2081–2092

Hashiloni-Dolev Y, Shkedi S (2010) The regulation of preimplantation genetic diagnosis for sibling donors in Israel, Germany and England: a comparative look at balancing risks and benefits. In: Birenbaum-Carmeli D, Carmeli YS (eds) Kin, gene, community: reproductive technologies among Jewish Israelis. Berghahn Press, Oxford, pp 61–83

Hawkins A, Ho A, Hayden M (2011) Lessons from predictive testing for Huntington disease: 25 years on. J Med Genet 48(10):649–650

Inthorn J, Wehling P, Schultz S et al (2014) Präkonzeptionelle Anlageträgertests: Diagnostik mit Fragezeichen. Dtsch Arztebl 111(9):A343–A346

Konrad M (2003) Predictive genetic testing and the making of the pre symptomatic person: prognostic moralities amongst Huntington's-affected families. Anthropol Med 10(1):23–49

Konrad M (2005) Narrating the new predictive genetics: ethics, ethnography and science. Cambridge University Press, Cambridge

Lemke T (2002) Genetic testing, eugenics, and risk. Crit Public Health 12(3):283–290

Lemke T (2005) Beyond genetic discrimination: problems and perspectives of a contested notion. Genomics Soc Policy 1(3):22–40

Lipkin M Jr, Rowley PT (eds) (1974) Genetic responsibility: on choosing our children's genes. Plenum Press, New York

Novas C, Rose N (2000) Genetic risk and the birth of the somatic individual. Economy and Society 29(4):485–513

Nozick R (1974) Anarchy, state, and utopia. Basic Books, New York

Raz A (2004) "Important to test, important to support": attitudes toward disability rights and prenatal diagnosis among leaders of support groups for genetic disorders in Israel. Soc Sci Med 59(9):1857–1866

Raz A (2005) The gene and the genie: tradition, medicalization, and genetic counseling in a Bedouin community. Carolina Academic Press, Durham

Raz A, Vizner Y (2008) Carrier matching and collective socialization in community genetics: Dor Yeshorim and the reinforcement of stigma. Soc Sci Med 67(9):1361–1369

Rose N (2006) The politics of life itself: biomedicine, power, and subjectivity in the twenty-first century. Princeton University Press, Princeton

Rosner G, Rosner S, Orr-Urtreger A (2009) Genetic testing in Israel: an overview. Annu Rev Genomics Hum Genet 10:175–192

Sandel MJ (2007) The case against perfection: ethics in the age of genetic engineering. Harvard University Press, Cambridge

Schroeder-Kurth T (1999) Screening in Germany. In: Chadwick R, Shickle D, ten Have H et al (eds) The ethics of genetic screening. Kluwer, Dordrecht, pp 81–89

Slowther A (2008) Predictive testing and population screening. Clin Ethics 3(1):11–13

Taylor S (2004) Predictive genetic test decisions for Huntington's disease: context, appraisal and new moral imperatives. Soc Sci Med 58(1):137–149

Vallance H, Ford J (2003) Carrier testing for autosomal-recessive disorders. Crit Rev Clin Lab Sci 40(4):473–497

Vardi-Saliternik R, Friedlander Y, Cohen T (2002) Consanguinity in a population sample of Israeli Muslim Arabs, Christian Arabs and Druze. Ann Hum Biol 29:422–431

Vehmas S (2001) Just ignore it? Parents and genetic information. Theor Med 22:473–484

Wertz DC, Fletcher JC (eds) (1989) Ethics and human genetics: a cross cultural perspective. Springer, Berlin

Zlotogora J (2010) The molecular basis of autosomal recessive diseases among the Arabs and Druze in Israel. Hum Genet 128(5):473–479

Zlotogora J (2015) Using community genetics for healthy consanguinity. In: Shaw A, Raz A (eds) Cousin marriages: between tradition, genetic risk and cultural change. Berghahn Press, Oxford, pp 175–184

Zlotogora J, Carmi R, Lev B et al (2009) A targeted population carrier screening program for severe and frequent genetic diseases in Israel. Eur J Human Genet 17(5):591–597

Chapter 5
Making Responsible Life Plans: Cultural Differences in Lay Attitudes in Germany and Israel Towards Predictive Genetic Testing for Late-Onset Diseases

"Que sera, sera—Whatever will be, will be—The future's not ours to see," so we are told by the famous oldie sung by Doris Day in, ironically enough, the movie titled *The Man Who Knew Too Much* (1956). What will life be like when we get old? Predictive genetic testing for late-onset diseases is an attempt to provide some partial answers to that unanswerable question. In cases of colorectal cancer that is running in the family, where an effective treatment exists, predictive genetic testing is recommended by medical experts even for minors (Duncan et al. 2008). However, with some late-onset diseases—such as Huntington's disease (HD) or dementia—being untreatable, the availability of predictive testing has re-opened the discussion concerning "the right not to know" (Erez et al. 2010).

In this chapter we focus on predictive genetic testing of adults for late-onset diseases. This domain of genetic testing is seen by providers as increasing the range of choices open to the healthcare consumer. A typical example is predictive genetic testing for breast and ovarian cancer of healthy, pre-symptomatic individuals (Levy-Lahad and Friedman 2007). Our research highlights the cultural diversity of lay (as opposed to expert) moralities concerning predictive testing, and how these public configurations interact with the top-down biomedical worldview of needs and benefits as well as with the Foucauldian approach to genetic testing as a uniform and unilateral regime of bio-governmentality (Novas and Rose 2000; Lemke 2002, 2005). This research also examines how public, inter-personal and subjective configurations of predictive testing reflect and interact with particular cultural repertoires (Raz and Schicktanz 2009a, b), as well as express forms of resistance, avoidance, and criticism of genetic testing. Lay deliberations of predictive genetic testing, involving discussions of responsibility for the self, self-care, and responsibility for family members, are used to highlight relevant moral grammars and their embedding of social norms and individual behaviors. We examine these diverse expressions by looking at arguments developed by lay people in Germany and Israel, examining the complex ways in which different cultural backgrounds, as well as experiences of being affected, influence the ways in

A.E. Raz and S. Schicktanz, *Comparative Empirical Bioethics: Dilemmas of Genetic Testing and Euthanasia in Israel and Germany*, SpringerBriefs in Ethics, DOI 10.1007/978-3-319-32733-4_5

which people make sense of predictive genetic testing in the context of making life plans.

The focus group discussions that were conducted in Israel and Germany included scenarios of testing for Huntington's disease and colon cancer, with questions about whether to test, when, and whether and whom to tell about test results. In the colon cancer scenario, participants in Germany and Israel were asked to imagine that one of their parents was diagnosed with colon cancer and that they are in their early 20th and consider taking a genetic test that can be performed at any stage of life, even before the onset of symptoms. Participants were told that the test will tell them whether they have a mutation that correlates with disease onset and which could, under specific environmental conditions (whose characteristics are still unknown) increase their probability to develop colon cancer later in life. It was also explained that at an early stage, colon cancer is often curable. Participants were then asked about their arguments for and against taking this genetic test, and about sharing test results with others.

Participants in Germany and Israel were also presented with a scenario asking them to imagine that one of their parents was diagnosed with HD, a severe neurological disease with degenerative symptoms starting around middle age. They were told that although the disorder itself is not fatal, complications such as pneumonia, heart disease, and physical injury from falls can reduce life expectancy to around twenty years after the onset of symptoms. It was explained that there is no cure for HD, and full-time care is required in the later stages of the disease. Participants were then told that a genetic test can be performed at any stage of life, even before the onset of symptoms, and that the chance for them is 50 % to have the gene. If they have it, it will, in virtually all cases, cause this disease to manifest itself in later life. Participants were asked about their arguments for and against taking this genetic test and sharing test-results with others. Quotes presented in the following section illustrate the range of responses with regard to two major themes that emerged from the analysis: self-knowledge and self-responsibility ("to know or not to know") and responsibility for kin ("to tell or not to tell").

5.1 To Know or not to Know?

A major argument among all the respondents in favor of knowing certain genetic risk factors was the availability of preventive or curative strategies:

> My ambition, I suppose like that of all people, is to live to 120 years and live a healthy life as much as possible. Because you can perform a surgery to cure the disease when detected early, I would rather perform the test as early as possible in order to undergo the operation in time, and not discover the disease when it's too late (Israeli female, lay group).

In Israel, the view in favor of testing in case of potential medical treatment was apparent in the case of colon cancer across all groups. It was expressed by the majority. While the majority of the modern-religious respondents were in favor of taking the test when medical treatment is available, they did not relate this attitude explicitly to their religious stance. The largest support of this view—that tests should be taken only when medical treatment is available or prevention is possible—was expressed in the two affected groups in Israel. In these groups, also the majority of the respondents also said that they will change their lifestyle and increase their medical surveillance as a result of a positive test:

> If I am found to have the gene I will perform the medical tests for early detection of colon cancer, to increase the chances of prevention or cure. I do not think there are any arguments against such a genetic test since if colon cancer is treated following early detection this may improve the effectiveness of prevention (Israeli male, affected group).

Only two Israeli respondents in each group were against taking the test in the context of colon cancer. They also mentioned the right not to know:

> I can still monitor my health without the genetic tests, can't I? This will still allow me to treat the disease in time if needed (Israeli female, affected group).

This minority view *against* taking the test even in a situation where treatment is available also contained references to the ambivalence resulting from the uncertainty of the expression of colon cancer even for carriers. A few Israeli respondents supported taking the test because "it contributes to the family," arguing that responsible parenthood meant that one had to take care of his/her health:

> I think that for your children you have a responsibility, your life is no longer just yours but also your commitment to your children and that's why I think every person, especially if he has children, needs to take the test if he belongs to a risk group. I have no objections to the test and I think that whoever raises these arguments is irresponsible (Israeli female, affected group).

In Germany, in contrast, only a minority was in support of taking the test, expressed in the context of colon cancer by less than a third of the participants. Arguments mentioned by many German respondents as part of the view against taking the test included the problem of probabilistic knowledge that is not helpful, anxiety, and the danger of discrimination: "I know from experience that if you know something like that, it will be a burden" (German female, lay group).

Another argument against taking the test stressed the existence of risk as a normal part of life that should not be medicalized:

> Well, I would also not take the test. In principle. Because I find that we must not live in a society where everyone can know what he or she gets, or will get. We should live in a society where we allow the existence of risks. Just like crossing the street (German female, lay group).

The largest support of taking the test for colon cancer was expressed in the affected German group:

> Well, I am very much in favor of early diagnosis because I learned much about cancer during my rehabilitation time. For three times, I was surrounded by 300 cancer patients. Then you'll know what early or late detection means. There is a great difference. I was diagnosed early and therefore it was curable. But someone who does not undergo cancer screening for five or even three years, even if he/she is a high risk patient… I find that, shall I say, somewhat negligent (German female, affected group).

In both Israel and Germany, the Huntington's disease scenario was approached with much more ambivalence, highlighting the type of information that is regarded as helpful or not in the context of predictive genetic testing:

> In this case [Huntington's disease], I'll skip the test. If there is no cure, I would not want to know - I'd rather live my life peacefully and without fear. The disease anyway bursts, and it's better not to know about it beforehand. I cannot find arguments for the test (Israeli female, lay group).

In Israel, the view in support of taking the test for Huntington's disease was expressed by a very heterogeneous range of the participants, depending on the group. Support was highest in the non-affected Israeli group and lowest in the modern-religious Israeli group. Many amongst those who were supporting this view in the non-affected lay groups argued in favor of knowing the result in order to be able to be better prepared in terms of life planning that is being conducted together with other family members:

> This question is very difficult. Yes, I would check. I want to prepare for the future, for my home, for family and financially. I think it's my responsibility as a mother to know in advance and prepare for it, responsibility towards my family… (Israeli female, lay group).

It was common amongst Israeli respondents, but not German respondents, to speak about their motivation for taking the Huntington's disease test in the context of their children, as one Israeli respondent summarized it: "[Taking the test] is my responsibility as a mother." Israeli respondents brought up prenatal testing in this context even though the moderator focused only on adult predictive testing. They argued that it is crucial to know about one's "faulty" genes so as to know which diseases to test for prenatally or avoid having children. The option of adoption was not mentioned by any of the Israeli respondents, and outspoken attitudes supporting selective abortion were not met with criticism, as the following exchange illustrates.

> Of course, if there is likelihood that I pass to my children any illness I would like to know about it, so I can monitor the fetus during pregnancy and have an abortion if necessary… As part of my responsibility as a parent, I am responsible to my child's health and this responsibility extends to the pre-fertilization stage (Israeli male, affected group).

Another group member responded to this by saying:

> I'm all in favor of taking the genetic tests, if the disease could pass to the fetus, I think it is right to do it, I would not want to have a child with an illness or disability, who would suffer throughout his life and be mistreated by society. Of course if it is some mild issue that it is possible to live with, there is no real problem and I would not rush to test, especially if testing could endanger me or cause any bodily harm to me (Israeli female, affected group).

Interestingly, the majority in the affected group was ambivalent or even against taking the test for Huntington's disease, referring in this context to the right not to know and to anxiety, since there is no treatment. Compared with the colon cancer case, this argument highlighted a view of self-responsibility to one's health/own body:

> Because there is no cure for the disease, in this case [Huntington's disease] I see no reason to do the test. I will have to live with the knowledge that my parents have the disease and hope that it will not attack me… The body will wait for something which may never take place. If there was a remedy, I would attach great importance to conducting the test, but since there is none – it does not contribute anything to me, and I think that such news could just make me more anxious (Israeli male, affected group).

A large majority of the respondents in the modern-religious group similarly spoke against taking the Huntington's disease test, mentioning the right not to know:

> I see no reason to check if I'm a carrier, if there is no medicine or vaccination that could reduce the confrontation, the struggle with the disease, or to prevent an outbreak. I have an instinct to say no, this is not something I would turn to religion about – and I also have a feeling that if you know, you would only become more stressed… so it's more relaxing not to think about it (Israeli female, modern-religious group).

Another argument raised by the modern-religious respondents was stigma, especially in the context of dating:

> Being HIV positive is much more stigmatizing. But being a carrier can also stigmatize you, especially if you are in the dating stage and have to talk about this (Israeli female, modern-religious group).

Another group member responded to this by saying:

> Yes that's true. So we should recommend to young people *not* to test for Huntington's - they can do this much later in life (Israeli male, modern-religious group).

In Germany, the view in support of taking the test for Huntington's disease was expressed by very few participants. Arguments given in favor of taking the test included the certainty of the genetic prediction as a potential source of comfort, and an important source of information for planning one's life:

> Well, Huntington's is an extreme case. Since there are no …treatments, it merely concerns knowledge. You have it, or you don't. I am not sure, but I believe that I would undergo a test in this case. Because, first, I know that my parents, or my mother has Huntington's. That means, that in any case the Sword of Damocles is hanging over my head, and I have a 50 % chance that it is this or that way. That means, the test is done, and there is a 50 % chance that I am well. I can continue living joyfully. And a 50 % chance that I have this disease. This then is indeed terrible. But at least I am aware of the situation. And I personally think that it is better for me if I know that I can still live ten or 15 years without a disability. And then, I believe, there are five more years of nerve disorder where one loses motor functions and the like. I think I can plan my life differently for this time if I know that I have actually still 50 years ahead (German male, lay group).

An additional argument in support of taking the test was raised by a few respondents mentioning that for them the question of whether or not to test would arise only in connection with the wish to have children, as we discuss below in more detail.

A slightly higher proportion of the German respondents were against taking the test for Huntington's disease. Arguments mentioned against taking the test included unnecessary anxiety:

> The question is whether I need this exact time specification to live a good life. Already if I know that this could probably be, I could try to live my current life as good as possible and as happy as possible. I don't have to know: In 15 years it will be my turn. I find that this time specification will drive one crazy. It is enough to know that this disease lies ahead of you. That is difficult enough to cope with. I don't have to assign a date to it (German female, lay group).

5.2 To Tell or not to Tell?

The formal right to know or not to know, as advocated by some lawyers and ethicists, tends to neglect the social implications of such knowledge. Individual disclosure of genetic status hence often leads to a new dilemma: whether one should convey this knowledge to relevant others. Therefore, it seemed rather self-evident for the participants to explore the social consequences of such knowledge. Interestingly, they spoke about this as a personal question of morality, or more explicitly of responsibility. However, the concept of responsibility varied according to different underlying norms such as love and care for others, duty and social obligation for those who are dependent, or self-identity as a family member. Even the decision not to tell could be framed as conveying responsibility to protect others from psychological harm.

Among our Israeli participants, considerable support for telling family members about the test results was apparent across all groups and was expressed in the case of colon cancer by a majority of more than half or even all participants. The largest support of this view was expressed in the modern-religious group, where 100 % of the respondents said they would share the test results with close family members, corresponding with the strong family-oriented characteristics of this population (Shalev et al. 2012). In the case of Huntington's disease, Israeli respondents' support was relatively high but also very heterogeneous. Support was especially high in the modern-religious group (almost all the participants said they would share test results with close family members). Although almost all of our respondents were against taking the test themselves, they still said that if they were to take it, they would share the results with others:

> In this situation there is no doubt that my children need to know. If you already have children you should tell them so they will have the opportunity to be tested themselves. I think it's a huge burden for a child and I would not tell them before I felt compelled to. When? I do not know, probably around age 17. On the one hand I know I want to save my child from this knowledge, on the other hand I think it's his right to know (Israeli female, affected group).

Arguments raised in the context of telling others about test results included responsibility for one's spouse and kin (sharing relevant information for life planning):

> I would tell my immediate family. I will definitely share it with my partner who needs to be aware of the risk and the possible impact of this on the continuation of our life together (Israeli male, lay group).

Telling kin was perceived as important because they need to be medically informed and take the test themselves:

> I would inform my immediate family, if I am found to be a carrier that is; they may be at higher risk now and so they need to decide for themselves if and how to take care of their bodies. They should be made aware that they are at higher risk than the general population (Israeli male, lay group).

Interestingly, the feeling of responsibility for their children was expressed also by taking the opposite stance: Some felt that they should share the test results only with their spouse and not with their children so as to not worry them too much:

> I would not tell my children so that not to make them worry. I think they have enough worries, I would not want to add (Israeli female, lay group).

In Germany, only a minority—in all groups—supported the idea of sharing test results of colon cancer with family members. A similar small proportion of respondents in the lay and modern-religious groups were against telling family members, and the majority either expressed ambivalence or did not voice an opinion regarding this issue. In the case of Huntington's disease, support of sharing results with family members dropped less than a third of the participants:

> Would then not the obligation exist to tell it to the others so they, too, can take this test? For example, to my own children? To suggest to them that they too could suffer from this genetic defect, or something similar. At least, to open up the possibility for them to take, or not to take, the test for themselves (German male, lay group).

Another respondent countered this view by expressing worries about telling others:

> Oh God! That is definitely 1984 for me. George Orwell. That is totally frightening and I am horrified. I will probably dream of that tonight… But if this would push through, you would take away the person's responsibility to keep this to himself [*sic*] because others might possibly be harmed by it. That is for me a sign that it is very difficult for us to accept uncertainty in life. There are things that simply happen. [In the past], I rode without a helmet. Now everybody rides wearing a helmet, and sometimes people look at me with disapproval if I don't wear one. I also ski without helmet. But there are countries where it is already forbidden to ski without a helmet. This means that I don't have this freedom anymore. And I am afraid that this will keep going on. On and on (German male, lay group).

German respondents who spoke against telling their family members emphasized the anxiety that such information may cause, and the unwillingness to share what was strongly perceived as personal information. Few German respondents discussed in this context of the decision not to have children, or—in case one carried a gene for Huntington's disease—the decision to adopt children:

> I also decided not to have children. But that has nothing to do with genetic testing. But let alone this: One has to deal with it. Not to have children is an inner effort. […] If I had a fifty-fifty, or let's say an eighty-twenty probability, then I would say: I refrain from having children: Then I'll adopt. I have five stepchildren. Then, I would rather adopt a child. To give the child a chance, instead of - sometimes religions see this somewhat differently - having my own child that would be severely disabled or something like that. I would probably blame myself throughout my entire life (German male, lay group).

Some German respondents also referred to the potential risk of being discriminated against by employers, friends and society. Once again, relatively the largest support of telling family members was expressed in the affected group:

> Well, I think if it's like that, it's good to be able to share that fear [of the disease] with somebody else (German female, affected group).

Moreover, some affected German respondents also spoke (in a parallel manner to the Israeli respondents) about the right to know of family members, which made them consider telling as their duty and part of their responsibility towards kin:

> Well, I'd consider it to be my duty. Yes. Well, I've always kept my relatives informed how my disease made its way through the genes of my family (German male, affected group).

The majority of Israeli respondents did not express worries regarding potential genetic discrimination, and many agreed that the State and HMOs should finance these tests as part of public health. For some, however, support for telling spouses and/or parents was accompanied by reservations concerning not telling children and the rest of the family:

> I would tell only my husband or my mother. I would not tell friends or even close relatives; I suppose they would still accept me as I am… But inside, they surely would treat me differently, feeling sorry for me and taking care of me more… I'm wondering if it's right in this case to tell the rest of the family that this is a terrible disease that accompanies fear for life (Israeli female, affected group).

Some German respondents who were in favor of telling family members also spoke about responsible sharing of information:

> We are somewhat supported by doctors or whomsoever, but our relatives will then be alone with me and this announcement (German female, affected group).

Other German respondents mentioned in this context that in addition to telling their family members about their test results, they would also inform them about relevant information centers where counselling is available. Some expressed mixed feelings, agreeing in principle that family members should be told but also sharing how, when they were told by their family members, this was for them a burden at the time because they were very worried about the results of the test they were going to take.

5.3 Conclusions

What can these findings tell us about the complex ways in which different cultural backgrounds, as well as the common experience of being affected, influence the ways in which people make sense of predictive genetic testing for their life plans? A major argument in favor of taking the test was the availability of preventive treatments. In Israel, this view was apparent across all groups and was expressed by the majority. In Germany, support of taking the predictive test even when a pre-emptive treatment exists was, in stark contrast, a minority view. This finding appears to reflect a significant cultural difference between lay people in the two countries. The Israeli respondents stressed the benefit of the test as leading to pre-emptive treatment; while the German respondents emphasized that the treatment is available irrespective of the person's knowledge about his/her genetic status, and so the test does not make a difference. A parallel finding reflecting cultural differences was apparent in relation to telling others. In Israel, support for telling family members about test results was expressed across all groups by the majority. In Germany, support of telling family members about the test results was a minority view in all of the groups. In the case of testing for Huntington's disease, the same picture emerged (as in the context of colon cancer) when comparing Germany and Israel, but with overall lower percentages of support for testing and for telling others, as a result of the lack of medical treatment.

The findings demonstrate how lay concepts of responsibility are used as complex vehicles for meaning and values deemed important for the participants in the context of predictive genetic testing. This finding is theoretically remarkable, substantiating theories about how (bio)medicalization implies a somatic responsibilization that is understood as a "regime of the self" (Rose 2006, p. 134). At the same time, this finding goes beyond this theorization in illustrating the need for further differentiation. Importantly, respondents seemed to draw upon different (implicit) understandings of responsibility involving multi-faceted configurations of various types of relationships (responsibility towards self, or others, or society at large), various types of temporal orientation (future-oriented or past-oriented), and various normative frames (rights and duties, for example). The theme of responsibility for oneself (self-responsibility) was mentioned by many Israeli respondents in the context of supporting testing but usually also with reference to one's family, namely that responsible parenthood meant taking care of your health also for the sake of your children. In contrast, German respondents mentioned self-responsibility as strongly connected to the perceived personal beneficence of the predictive test, in terms of the difference that knowing the genetic information can make for the individual. They also pointed out in this context concerns about the medicalization of risk that should be seen as a natural part of the human condition.

These differences are arguably connected to broader cultural scripts. Self-responsibility, in the Israeli groups, was understood as being primarily the responsibility to stay healthy (for people's own needs as well as in the context of their commitments to family members). In Germany, self-responsibility also meant to accept risk as part of life, and therefore to organize one's life around these risks—including the social acceptance of diversity and illness. This also had to do with how much certainty and agency participants wanted, and how they integrated predictive testing into this. Israeli respondents prioritized health care, and were anxious about possible illnesses, while German respondents argued that probabilities are not helpful and that society should not be built on certain ideas of health control.

In case they were found to carry a severe genetic disease, the desire to have healthy children led many Israeli respondents to support testing the fetus; whereas German respondents spoke about deciding against having children or otherwise preferring adoption. This mirrors procreative versus individualistic discoursive patterns, common in reproductive issues. For participants in both countries, their hypothetical life plan in such a situation was connected with what they perceived as responsibility, but interestingly, with totally different implications. Such dialectical focus on the cultural grammars behind individual and interpersonal concepts of responsibility provides a helpful, although little explored interface, for bridging some of the gaps between experts' formal ethics of principles and our lay moralities, and between theoretical and empirical bioethical analysis (Schicktanz and Raz 2012).

The conceptualization of responsibility in lay moralities that emerges from this study is not all personal and idiosyncratic but rather framed by broader socio-cultural and ethical narratives, as already mentioned. This more nuanced, culture-based understanding of responsibilities can be used to flesh out a concept that so far has remained very abstract, for example in the work of many communitarians, and develop this concept further on an empirical level. For philosophers of ethics like Baylis et al. (2008), concrete expressions of relational solidarity in the context of public health ethics are to be found in our accepting of responsibility for ourselves and our actions; in our willingness to be held accountable for others (especially the weakest and most disadvantaged in society); and in our awareness of mutual vulnerability and interdependence. However, because responsibility is always about our relationships to others, it is neither purely communitarian nor strictly liberal, because both concepts rather describe the relationship of citizens to the state and vice versa (Kymlicka 1990). It rather occupies a third, hybrid space of morality in-between these two opposites, as it concerns the social life of families, clans, and collectives below the state level.

Our findings highlight and contextualize these different and hybrid meanings of responsibility as embedded in cultural grammars of individualism/collectivism as well as acceptance/rejection of disability and illness as part of social diversity and social life. In the case of Israeli respondents we see a relational responsibility blurring the boundaries between the individual and his/her family, reflecting a more traditional, family-oriented, close-knit society that highlights the importance of genetic kinship ties (Birenbaum-Carmeli 2010; Hashiloni-Dolev and Shkedi 2007). In the case of German respondents, we see a self-responsibility based on self-determination, reaffirming the boundaries between the individual and his/her family, reflecting a more liberal-oriented, loose-knit society that highlights the importance of social parenting (Wiesemann 2010; Hashiloni-Dolev and Shkedi 2007).

Being affected, in contrast to the abovementioned cultural grammars, was found to be connected with a line of arguments that in some cases provided more uniformity. In the case of testing for colon cancer, the largest support for knowing and for telling (the common view amongst affected as well as non-affected respondents in Israel) was expressed in the German *affected* group, highlighting the role of "being affected" in producing a phenomenological source of uniformity that transcends national diversity. In Israel, as other studies have shown, "genetic anxiety" (or "responsibility," depending on one's perspective) is constructed by professionals, experts and the public to provide a collective frame of risk in which the "elective" uptake of genetic testing is exceptionally high and seen by many as moral duty (Remennick 2006). The moral argument regarding the duty to know reflects, in the case of Israel, a confluence of public and professional (medico-legal) worldviews promoting genetic testing as a collective agenda (Zlotogora et al. 2009). The moral argument regarding the right not to know due to anxiety and stress, supported by the German non-affected lay groups (but not by many in the affected groups) may also reflect a romantic tradition of scientific skepticism as well as a dystopian view of medicine and criticism of genetic testing (Hashiloni-Dolev 2007). Overall, this study demonstrates that a multifaceted awareness of the variety of worldviews, including the attitudes of those affected by clinical symptoms or by being a carrier, and of those not affected, warrants attention by sociologists, bioethicists and all those who are interested in a more nuanced understanding of genetics as social practice.

References

Baylis F, Kenny NP, Sherwin S (2008) A relational account of public health ethics. Public Health Ethics 1(3):196–209

Birenbaum-Carmeli D (2010) Genetic relatedness and family formation in Israel: lay perceptions in the light of state policy. New Genet Soc 29(1):73–85

Duncan RE, Gillam L, Savulescu J et al (2008) "You're one of us now": Young people describe their experiences of predictive genetic testing for Huntington disease (HD) and familial adenomatous polyposis (FAP). Am J Med Genet Part C 148C:47–55

Erez A, Plunkett K, Sutton VR et al (2010) The right to ignore genetic status of late onset genetic disease in the genomic era; prenatal testing for Huntington disease as a paradigm. Am J Med Genet Part A 152A:1774–1780

Hashiloni-Dolev Y (2007) A life (un)worthy of living: reproductive genetics in Israel and Germany. Springer, Dordrecht

Hashiloni-Dolev Y, Shkedi S (2007) New reproductive technologies and family ethics: pre-implantation genetic diagnosis for sibling donor in Israel and Germany. Soc Sci Med 65(10):2081–2092

Kymlicka W (1990) Contemporary political philosophy: an introduction. Oxford University Press, Oxford

Lemke T (2002) Genetic testing, eugenics, and risk. Critical Public Health 12(3):283–290

Lemke T (2005) Beyond genetic discrimination: problems and perspectives of a contested notion. Genomics Soc Policy 1(3):22–40
Levy-Lahad E, Friedman F (2007) Cancer risks among BRCA1 and BRCA2 mutation carriers. Br J Cancer 96(1):11–15
Novas C, Rose N (2000) Genetic risk and the birth of the somatic individual. Econ Soc 29(4): 485–513
Raz A, Schicktanz S (2009a) Lay perceptions of genetic testing in Germany and Israel: the interplay of national culture and individual experience. New Genet Soc 28(4):401–414
Raz A, Schicktanz S (2009b) Diversity and uniformity in genetic responsibility: moral attitudes of patients, relatives and lay people in Germany and Israel. Med Health Care Philos 12(4): 433–442
Remennick L (2006) The quest for the perfect baby: why do Israeli women seek prenatal genetic testing? Sociol Health Illn 28(1):21–53
Rose N (2006) The politics of life itself: biomedicine, power, and subjectivity in the twenty-first century. Princeton University Press, Princeton
Schicktanz S, Raz A (2012) Responsibility revisited. Med Stud 3(3):129–130
Shalev O, Baum N, Itzhaky H (2012) Mate selection and marriage decision in bicultural society: modern orthodox society in Israel. Marriage Fam Rev 48(2):210–226
Wiesemann C (2010) The moral challenge of natality: towards a post-traditional concept of family and privacy in repro-genetics. New Genet Soc 29(1):61–71
Zlotogora J, Carmi R, Lev B et al (2009) A targeted population carrier screening program for severe and frequent genetic diseases in Israel. Eur J Hum Genet 17(5):591–597

Chapter 6
Planning One's End of Life in an Expert Biomedical Culture

A comparison of Germany and Israel can highlight the complex impact of cultural factors on the moral discourse surrounding end-of-life decisions. The comparison shows interesting differences in how patients' autonomy and doctors' duties are morally and legally related to each other with respect to the withholding and withdrawing of medical treatment in end-of-life situations. Taking the statements of the two national expert ethics committees on end-of-life in Israel and Germany as well as their legal outcome as an example of this discourse, we first describe their similar recommendations. Then we focus on the differences, including balancing ethical principles, what is identified as a problem, what social role professionals play, and the influence of history and religion. The comparison shows Israel to be more restrictive in relation to Germany, contrasting previous bioethical studies in which Germany was characterized as much more restrictive, for example in the context of the moral and legal discourse regarding the beginning of life. We conclude by reflecting on the ambivalence of the cultural reasons for this difference and its expression in various dissenting views on passive euthanasia and advance directives.

6.1 Recent Changes in the Legal Situation

In both countries, the recommendations of interdisciplinary expert committees appointed by the government paved the way to ensuing legislation: the report of the German National Ethics Council on self-determination and care in end-of-life decisions (National Ethics Council 2006) and the report of the Public Committee on matters concerning the terminally ill patient (Public Committee 2006) in Israel. A comparison of the two documents shows that there is a consensus about the framework within which end-of-life decisions should take place: providing basic care for all patients on the one hand and limiting patient self-determination to cases of passive euthanasia on the other hand (Schicktanz et al. 2010a). Besides these

A.E. Raz and S. Schicktanz, *Comparative Empirical Bioethics: Dilemmas of Genetic Testing and Euthanasia in Israel and Germany*, SpringerBriefs in Ethics, DOI 10.1007/978-3-319-32733-4_6

similarities, the reports also show significant differences based on religious and cultural factors (see below) that are mirrored in the legislation (for an in-depth comparison see Schicktanz et al. 2010b).

In both countries voices of religious groups played an important role in the debate on end-of-life decisions (Shapira 2006). In Germany, these included the Roman Catholic Church (Kongregation für die Glaubenslehre 1980), and Protestant churches (Frieß 2008; Schardien 2007). Both churches agree that active euthanasia should not be legal, but pain relief and all forms of withholding and withdrawing treatment in accordance with a patient's will are seen positively.

The Orthodox Jewish (halakhic) tradition, in contrast, views the sanctity of life as more important than self-determination. Withdrawing treatment is considered as an active, thus unacceptable, life-shortening intervention. Withholding treatment, however (for example by not resuming treatment), is viewed as passive and thus considered positively under certain circumstances (Barilan 2004, 2012; Schicktanz et al. 2010b).

These differences are mirrored in the legal situation in Germany and Israel. In Germany, the law on advance directives (2009), which is part of the guardianship legislation,[1] places the emphasis on death and end-of-life decisions as primarily an individualized event. It shows a broad acceptance of a variety of advance directives as expressions of patient's autonomy (Jacobi et al. 2005; Wiesing et al. 2010). In Israel, the Dying Patient Act (2006) is a stand-alone law dealing with advance directives. It frames death and end-of-life care as primarily social events, defined by Jewish halakhic restrictions on autonomy, with the application of advance directives allowed, but only on a very restrictive and bureaucratic basis (Barilan 2004, 2012; Jotkowitz and Glick 2009; Shalev 2009, 2010). Overall, the Israeli law has two major unique attributes: first, the application of advance directives is limited to terminally-ill patients in the last 6 months of their life; second, only withholding treatment is allowed. Israeli advance directives are provided on a long, jargon-loaded formal form issued by the Ministry of Health. The German legal position is rather liberal. An advance directive must be respected in any decision concerning medical treatment, regardless of the stage of the illness. In Germany, there are different advance directive forms issued by various agencies: the Ministry of Health, medical associations, social movements, patient advocacy groups and the Churches, that are all legally binding.

A new development in Israel concerns the legalization of physician-assisted suicide. On June 8, 2014, a bill proposed by MK Ofer Shelah legalizing physician-assisted suicide passed the Knesset's Ministerial Committee for Legislative Affairs. If passed, the new law would join the existing Law for the treatment of terminally ill patients, allowing doctors to administer a lethal injection to patients who have 6 months or less to live. The proposed model follows that of the U.S. State of Oregon,

[1]See Drittes Gesetz zur Änderung des Betreuungsrechts, in: Bundesgesetzblatt 2009 I, 48:2286 −2287. http://www.bgbl.de/xaver/bgbl/start.xav?startbk=Bundesanzeiger_BGBl&jumpTo=bgbl109s2286.pdf. Accessed 03 Aug 2015.

in which two doctors must confirm a diagnosis of terminal illness. The law includes various safety precautions to ensure that the patient is in a clear mind and acting with self-determination. It is required that the patient requests the prescription for a lethal dose of medicine three times within a 2 week-period gap between each request; the patient must be a local resident for at least 5 years, they must be of legal age, and only then will they be able to receive the prescription; they must also pick it up on his own.

Stakeholders' views on the new bill diverge, with opposition mainly from Orthodox circles and the Israel Medical Association. Chairman of the Israel Medical Association, Dr. Leonid Eidelman, argued that while

> We feel empathy for the severe suffering of terminal patients and their families, but it is not the role of the doctor to end the life of a terminally ill patient in a way that isn't natural. We are not trained to do so and do not teach the next generation of the medical community to do so. Euthanasia and suicide assistance with the help of a doctor are unethical actions.

The new bill is also firmly opposed by religious leaders. Chief Ashkenazi Rabbi David Lau expressed his opposition, saying that "a doctor can be a healer, but when he is unable to heal he does not have the authority to kill. He may give pain medication, even in heavy doses, as death nears, but cannot kill." And in the Jerusalem Post Nadel (2014) wrote:

> Israel is not Belgium, the Netherlands, or the State of Oregon. The State of Israel is the Jewish state, and religious affiliations aside, Judaism values life above all. In Judaism, life has sanctity. A law legalizing physician-assisted suicide threatens to undermine the Jewish character of the Jewish state.

Lilach, the Israel Association for the Right to Live and Die with Dignity, welcomed the decision:

> The bill will give terminally ill patients for the first time the decision to end their lives whenever they choose to do so, and not wait for death with suffering and anguish. Today, their only option is euthanasia in Switzerland, which is very pricey and not an option for most patients. Contrary to claims that the proposal can be a slippery slope to suicide, the reality has proved the opposite. We call upon members of the Knesset to support the bill and allow patients to end their lives with dignity.

Notably, in Germany an ongoing debate was initiated by a few ethicists and politicians in 2014 by suggesting a particular legal regulation for physician assisted suicide. While one proposes to forbid it explicitly, others plea for allowing exceptions from this ban but only after strict proof of the conditions. A third voice occurred now to argue for no regulation. This position supported by representatives of the Christian churches, was perhaps triggered by the fear that the majority during parliament discussions might rather lead to a very permissive final solution. The parallel timing of such a discursive event in both Israel and Germany can be understood as framed by a rather general tendency in many modern Western countries (e.g., Netherlands, Belgium, and Switzerland). The debate between the various stakeholders, including political parties, Orthodox leaders, and professional and self-help organizations, will be traced in more detailed in the following section.

6.2 Tracing the Differences Between Germany and Israel in End-of-Life Care Policies

In 2006, the former Israeli prime minister Ariel Sharon suffered a massive stroke and fell into coma. Subsequently, he received life-sustaining treatment and was being kept in a pervasive vegetative state until he died in 2014. The media coverage of his case illustrates the impressive interests as well as opens debates concerning life-saving and life-prolonging treatments for patients who remain unconscious for years. A recent report from Germany includes a quote blaming his sons for keeping their father alive out of selfishness (Inbari and Gil 2013). In a recent NBC news report, medical ethicist Arthur Caplan claimed that "[k]eeping Sharon or others like him alive in a very damaged, extremely limited state with no hope of recovery is not something that the government should pay for without some support from those who want life to go on" (Caplan 2013). In Israel, where the withdrawal of artificial nutrition is legally prohibited, media comments on the comatose Prime Minister have been relatively non-controversial, with some recent newspaper excitement concerning alleged brain activity. In a 2010 art exhibition in Tel-Aviv, a life-sized sculpture of Sharon lying in a hospital bed was put on display. The gallery curator described the exhibit as "an allegory for the Israeli political body—a dependent and mediated existence, self-perpetuated artificially and out of inertia, with open eyes that cannot see." (Simon 2010). Such end-of-life dilemmas are becoming more wide-spread and common for lay people. They are influenced by cultural and religious factors and take place within different national and medico-legal frameworks. In this study, the juxtaposition of Israel and Germany serves as an example of two opposite regulatory approaches to end-of-life decisions. While the medico-legal policies and institutional responsibility for end-of-life care and advance directives are well discussed (Schicktanz et al. 2010a, b), less attention has been given to how lay people—and particularly people who are affected by end-of-life care issues—perceive and frame such responsibilities. Therefore, we examine the attitudes of German and Israeli lay people, including modern religious people and those directly affected (e.g., relatives of patients) toward issues of end-of-life care.

End-of-life decisions concerning passive and active euthanasia, or handling advance directives, are very complex and closely linked to legal, medical, religious, and bioethical discourses. We want to examine and deconstruct these linkages in Germany and Israel, moving beyond one-dimensional constructions of ethical statements as "social facts" to their conflicting and multi-faceted embedding within professional, religious, and cultural perspectives (Turner 2005). So far, bioethical comparisons between Germany and Israel have generally shown Germany to be more restrictive and Israel to be more permissive. Various studies have shown that German and Israeli professionals such as genetic counsellors, bioethicists and physicians differ strongly in their moral assessment of various biomedical issues including pre-implantation and prenatal genetic diagnosis, stem cell research and issues of selective abortion (Wertz and Fletcher 1989; Hashiloni-Dolev 2007), Gottweis and Prainsack 2006). While all these studies have dealt with the beginning of life, here we consider the other side of the lifespan, namely end-of-life decisions. We want to explore whether the same moral opposition also characterizes end-of-life ethics in the two countries. In the moral assessment and evaluation of end-of-life dilemmas. We begin by focusing on the variation in the expert, declarative discourse of policy and regulation. Especially in the German end-of-life debate, the specter of the "Nazi doctors" has played a significant argumentative role (see, e.g., Schmuhl 2000; Kröner 2001; contrarily arguing: Ach and Gaidt 2000). It is worth to mention, that the term "euthanasia" has in Germany difficulties to be accepted. The German National Ethics Council introduces explicitly the term "criminal euthanasia" (2006) to classify this murdering during Nazi time under the more general label of "euthanasia." The descriptive part of this study presents an ethical analysis of the statements of two national expert committees on end-of-life in Israel and Germany. This analysis leads us to discuss the notion of restrictive versus permissive bioethics, further embedding it in the cultural and religious context of assimilating medico-technical developments.

Our bi-national analysis of the end-of-life ethics discourse focuses on the official statements of two national expert committees in Israel and Germany both leading the current status quo. We analyzed these statements with a special focus on passive euthanasia and advance directives along the following research questions: (A) What are the main ethical principles mentioned? (B) What are the main problems identified? (C) What role do cultural and historical factors play? Furthermore, by deconstructing each of the statements and exposing their basic elements, which are then compared cross-culturally, this methodology aims at uncovering the embedding of ethics within multi-faceted and conflicting cultural narratives. In such manner we offer a subversive analysis of ethical declarations as presenting an illusion of being monolithic and consensual "social facts" (Haimes and Williams 2007).

A comparative investigation of the ethical debate on end-of-life decisions in Israel and Germany has not been done before. The two national ethics committees were both found in a politically powerful position. We compare on the one hand the 2002 Israeli Report of the *Public Committee* on "Matters Concerning the Terminally Ill Patient" (Public Committee 2006; the so called *Steinberg-Committee*, in the following abbreviated as IL-Report-DP) with the 2006 German *National Ethics Council's* Opinion Report on "Self-determination and Care at the End of Life" (in the following abbreviated as GE-Opinion-EOL) and the 2005 German *National Ethics Council's* Opinion Report on "Advance directives" (in the following abbreviated as GE-Opinion-AD). For the purpose of ethical analysis we focus primarily on the original and full statement of the expert committee, rather than on the latest version of the law itself (although an overview of the updated legal situation is provided), since the committee's statement contains the discussion of the premises and arguments underlying the formulation of the law (see also Lepping 2007). In Germany, on 1st of September of 2009 a new law was coming into force. According to the new law existing advance directives are obliged for attending physicians and strengthen the role of proxies in end-of-life decisions, too. A rather short, but very controversial parliamentary debate preceded this regulation. In Israel, a new law based on the expert committee report (and in line with most of its important issues) was promulgated by the Israeli Knesset in 2005; however, many practical questions concerning the implementation of the law are still under-regulated and debated.

The German National Ethics Council was appointed by the German Chancellor Gerhard Schröder in 2001 with the general mandate to discuss urgent bioethical issues and to advise the Chancellor, the parliament and the public. The Israeli committee was appointed by the Minister of Health Shlomo Benizri in 2000 with the specific mandate to propose legislation on end-of-life medical care.

Both committees were comprised of an interdisciplinary membership. At the time of its end-of-life report, the German Committee consisted of 25 members (professional philosophers, lawyers, medical scientists, patient representatives, politicians and religious authorities (Catholics and Protestants) and chaired by the lawyer and politician Christiane Weber-Hassemer. The Israeli Committee consisted of 59 expert members, making it the largest public committee ever established in Israel, and chaired by the pediatric neurologist and Jewish medical ethicist Steinberg (2006, p. 100). Both national committees aimed for a formal consensus statement which was achieved in the main parts, but both also offer some dissenting statements.

6.3 Comparative Analysis of the Statements

6.3.1 Legal Backgrounds

In both countries, to the date of the reports no specific law existed. However, in Israel, there were Ministry of Health guidelines (1996) and the Patient's Rights Law (1996), and several court decisions that indicated recognition of certain legal principles (e.g., the patient's right to refuse treatment, the prohibition of active euthanasia, *in dubio pro vitae*) (Shalev 2000). In Germany, there existed sections in the criminal law which were relevant to end-of-life decisions, as well as articles in the Constitution, such as Article 2(1), which protects the right to free personal development, and Article 2(2), which protects the individual right to life and physical integrity and was construed to include autonomous decisions in shaping the dying process. However, there remained considerable legal questions about the doctor's position as a guarantor of life, and a large degree of ambivalence regarding the doctor's duties.

6.3.2 Main Similarities and Differences

The main recommendations in both expert statements can be summarized with respect to their similar articulation of the following five points:

- There is a duty to administer palliative care.
- There is a duty to administer basic care, which includes hygiene, social care, and treatment to assuage feelings of starving and of thirst.
- Killing on request (so-called active euthanasia) should be legally forbidden. Both statements finally state that no doctors or other persons should be allowed to kill a patient (e.g., by applying a deadly dose of a substance), even on the patient's explicit wish.
- In case of doubt or uncertainty about the patient's wish, a decision in favor of life ought to be made: In the absence of an explicit statement or advance directive by the patient, and if the relatives seem to be unclear or contradictory as to the patient's presumed wishes, physicians and care givers should save life and not stop life sustaining care.
- The role of advance directives has to be strengthened and made legally binding.

The two reports differ in some major respects. The first discrepancy concerns the moral acceptance of *letting die* (a term suggested by the German NEC to be used for "passive euthanasia"). In the German report a strong consensus was expressed that "letting die" encompasses both withholding *and* withdrawing medical treatment. As long as letting die is in accordance with the patient's explicit wish, the NEC supported it. In contrast, the Israeli report stand out from most international reports and laws by stressing that there is a significant distinction between interrupting (withdrawing) and refraining from (withholding) medical treatment because of psychological, philosophical and religious (halakhic) reasons associated with the former. Hence, the report concluded, refraining from medical treatment is morally acceptable if in accordance with the patient's explicit voiced wish. This includes the withholding of any kind of medical treatment. However, interrupting a continuous treatment such as artificial respiration or artificial feeding is seen as morally unacceptable even if the patient wishes it (see the more detailed discussion below).

With respect to advance directives, the Israeli report stressed the need for a national data bank on advance directives and for medical caregivers to prompt patients to formulate an advance directive. Furthermore, the recent law defines the role of the doctor or the nurse not to testify to the validity of the AD, but to provide all relevant medical information. The NEC recommended making advance directives legally binding for physicians, care givers and legal representatives (GE-Opinion-AD) but did not address the practical problem of implementing advance directives in medical practice given the fact that only very few Germans hold an advance directive. The recent political discussion in Germany has resulted in four different motions in the parliament. Three suggest a law but differing in the regulation of how liable written (or even oral) advance directives should be. None of the motions has forseen a national data bank. The fourth motion has suggested not regulating advance directives at all.

6.4 Cultural Diversity of Moral Key Elements in the Argumentation

It is noteworthy that balancing various, sometime opposing values seems to be the key element in end-of-life decisions according to both committees' statements. Value balancing entails a twofold problem: First, one has to justify the main ethical principles; secondly, some hierarchy between the principles has to be considered. In the Israeli report four main ethical principles were mentioned: sanctity of life, prevention of significant suffering, quality of life, and patient autonomy. The German report referred to self-determination and protecting bodily integrity as part of patient autonomy, then to protection of life, and solidarity with vulnerable persons. However, "protection of life" as opposed to "sanctity of life" has a less religious and less absolute connotation. Furthermore, the German committee understood the right of self-determination as "the" ethos of contemporary life. In contrast, the Israeli report started with a general presumption that autonomy, as "the" democratic value, has to be balanced with other, Jewish religious values—of which "sanctity of life" is the most important one. This can be demonstrated by the non-liberal (socially prescriptive) recommendation of the Israeli committee that caregivers have a duty to persuade patients to accept oxygen, food, drink and regular medicine and not to withdraw medical treatment (IL-Report-DP, p. 221).

6.4.1 Handling Conflicting Situations in End-of-Life Situations

The role of experts represents a key element in both committees. The Israeli report delegated the responsibility for conflict resolution to medical and health care experts. According to the recent Israeli Dying Patient Law, institutional ethics expert committees decide in conflicting situations between the involved parties. In the case of unresolved dissent in the institutional ethics committee, a national committee is authorized to decide, by taking into account the patient's wishes according to his world view and lifestyle (see also: IL-Report-DP, p. 232). Concerning the same issue, a debate developed in the German committee. The German report generally remained imprecise regarding who should balance the norms and values: the physician, the patient, or the state; yet the answer can be found in the report of German NEC on advance directives: "in the event of conflict—in particular concerning the interpretation of the advance directive—the Court of Guardianship should decide." (GE-Opinion-AD, p. 57). Thus, the recent law took up exactly this reasoning by regulating that the Court of Guardianship has to decide when there are conflicts between physician and proxy with regard to serious life-shorting decisions. The cultural assessment of experts' responsibility is hence quite different: While in Israel health care experts are seen as best equipped to solve end-of-life dilemmas, in Germany more power is put in the hands of legal experts.

6.4.2 Religious and Historical Factors

The debate in the Israeli committee regarding the moral distinction between withholding and withdrawing treatment (passive and active "letting die") was also a fierce controversy in the international bioethics community, especially until the mid-1990s (see Callahan 1989; Steinbock and Norcross 1994; overview by Howard-Snyder 2011). However, most international statements on passive euthanasia nowadays agree that there is no substantial moral or legal distinction between the two because of consequentialist and intentional considerations. Correspondingly, the German Report emphasized that there is no morally relevant distinction between removing of a continuing medical treatment or not applying it (GE-Opinion-EOL, p. 46). The examples mentioned are artificial feeding and respiration. This NEC report concluded that the morally relevant action is to omit an unwanted treatment (p. 47), although acknowledging that it is "psychologically understandable if the doctor … is more reluctant to intervene actively—by removing a feeding tube or turning off a respirator—than simply do nothing" (ibid.). According to them, intention *and* outcome of the action are both morally relevant, so a moral distinction of actively doing and omitting was seen as inherently unjustified.

In contrast, the moral distinction between withholding and withdrawing played an important role in the Israeli report. The supporting argument is explained by the Chair Steinberg (Steinberg 2006, p. 103):

> According to relevant halakhic principles and the actual rulings of these prominent rabbis, preserving life is one of the most important values, but it is not an infinitive or an absolute value. Any act that shortens life is halakhically considered as murder, even at the very end of life (…). Therefore, active euthanasia, physician-assisted suicide or the withdrawal of a continuous treatment such as a pacemaker or a respirator is absolutely forbidden.

Interestingly, this argumentation ends in a novel distinction on the level of technology (and not on the level of action theory between doing and allowing)—a distinction between "continuous" and "discontinuous, serial" treatment. This distinction is based on the idea that medical treatment as such can be seen as one "act" or a sequence of acts: for example, the administration of dialysis is seen as a sequence of single, equivalent acts, while artificial respiration over several months is seen as only one act.

However, this moral distinction was criticized by various (secular) ethicists and lawyers in the Israeli committee as unethical and inconsistent with the previous case law (see Ravitsky 2006). Judges, legal scholars and philosophers maintained that there was no logical difference between such omission or action, in accordance with the international agreement. From a patient rights perspective, it was argued that just as informed consent is required before the administration of medical treatment, so too treatment may not be continued against the will of the patient, since in both cases the forced treatment is a violation of the right to bodily integrity and amounts technically to assault. Moreover, a rule forbidding the withdrawal of "continuous" treatment at the patient's request might be detrimental, because it could induce a patient to forgo potentially beneficial treatment out of fear that once it is begun it cannot be stopped. In addition, artificial respiration is often initiated as a matter of course in conditions of medical emergency, where the patient does not have a real option to indicate consent due to the circumstances. The rule of first treating and then asking questions in emergencies is entirely appropriate, but if discontinuation of treatment is not allowed, patients can find themselves held captive by life-prolonging measures which they might have preferred to refuse (see also Shalev 2009).

A technological solution was suggested to bridge the moral distinction that was erected between continuous and serial treatment. The solution suggested was the installation of a timer on a respiratory machine, which automatically switches off the respirator every day or so, and has to be intentionally turned on by someone again. This resetting of the clock would be routine care unless the patient explicitly expresses her wish not to turn it on again. The idea of a clock being reset, and thereby making continuous treatment into non-continuous treatment might look odd to an outsider. But it would look a little less odd if the readers recall the pervasive use of timers in orthodox households to get around Sabbath restrictions on lighting fires. The timer may be very useful for overcoming the understandable reluctance to "pull the plug," but it raises various questions. For example, if it is morally wrong to discontinue life support, what difference does it make *how* we perform the act? Secular lawyers and ethicists also asked whether a change of technique can make right of something that is wrong.

Concerning the question of historical considerations, this came up only in the German NEC, when its members discussed the ethical and social dimension of killing on request. The part of the members of the NEC considering killing on request as ethically acceptable (GE-Opinion-EOL, p. 86f) nevertheless supported the recommendation that it should be kept legally forbidden because of political considerations. These considerations might be interpreted as taking historical responsibility for criminal euthanasia practiced under the Nazi regime—an argument explicitly mentioned by members of the NEC. In the Israeli report we cannot find any comparable expression or reference.

6.5 Conclusions

In contrast to its general permissiveness in the beginning-of-life discourse, the Israeli view on end-of-life was found to be much more restrictive. This contrast between the beginning and end of life could draw on the "two-fold" view of Israeli society (Raz 2004), namely that in Israel there is a gradual separation between the pre-born (partial moral status of the embryo) and the already-born (full moral status), while in Germany full moral status is already attributed to the human embryo from the moment of conception. In addition, the relative permissiveness of Germany in the context of end-of-life can be seen to represent different social conceptions of the doctor's duty: with a German emphasis on doctors' duty to respect the self-determination of patient's and an Israeli focus on doctors' duty to respect the sanctity of life.

We would like to refine our own generalization by offering a more differentiated understanding of the term "restrictive." First, the Israeli stance concerning passive euthanasia is perhaps superficially more restrictive because of two reasons: (a) patients should be persuaded to accept life-supporting machines and artificial feeding, and (b) patients' expressed wishes (e.g., even written in an advance directive) for withdrawing a life-supporting-machine are—in a paternalistic manner—not accepted. However, regarding the practical solution of the timer, one could ask whether this is still considered "restrictive." It depends who will have the power to decide whether the timer is set up again—the patient (Barilan 2004) or the physicians and caregivers (as the committee has suggested).

The practical solution of the timer was regarded by some authors as a fig leaf for existing moral problems or even a "sell-out" (Butcher 2005; Ravitsky 2005). Instead, one could see this as a pragmatic way out for allowing individual solutions for existing cultural conflicts. In terms of Israeli culture, this pragmatism is a main cultural interface between Jewish tradition and secular modernism (Gross and Ravitsky 2003; Shapira 2006).

The Israeli position is sometimes explained by the fact that "sanctity of life" is seen as the highest religious (Jewish) ethical principle (Glick 1999; Green 1999; Ganz et al. 2006). Interestingly, Jewish ethics scholars such as Leonard Kravitz (2006) claim that the statements in the Steinberg committee's report do not present any universalistic position of Jewish thinking (especially with respect to the proscription of withdrawing treatment). In the German debate, too, the religious impact should not be underestimated. Christian authorities refer similarly to the sanctity of life and the moral unacceptability of terminating human life, seen as a gift from God (GE-Opinion-EOL, p. 75). However, the religious position is presented as one voice among many others, in a concert of pluralistic discourse. Moreover, representative surveys show that a majority of the German population is in favor of active euthanasia (GE-opinion-EOL). Whether the political class' as well as most of experts' reluctance is because of holding continuing loyalties to the opinions of the two Christian churches would be worth to be investigated in more detail. In addition, Israel is a democracy with a legal system in which patient autonomy plays an important role, so that the Steinberg committee also recommended to increase the relevance and binding character of advance directives. In contrast, the call of the German NEC in favor of strengthening advance directives was criticized by other interdisciplinary ethics committees [e.g., the German Interim Parliament Commission (so called Enquete-Kommission) for Ethics and Law of Modern Medicine]. In addition, the recent situation surrounding the question whether advance directives should be legally binding even if doctors and proxies disagree shows how patients' autonomy in Germany is still counterbalanced in practice with the decision making powers of doctors and judges.

References

Ach JS, Gaidt A (2000) Wehret den Anfängen? Anmerkungen zum Argument der „schiefen Ebene" in der gegenwärtigen Euthanasie-Debatte. In: Frewer A, Eickhoff C (eds) Euthanasie und die aktuelle Sterbehilfe-Debatte. Die historischen Hintergründe medizinischer Ethik, Campus, Frankfurt, pp 424–447

Barilan YM (2004) Is the clock ticking for terminally ill patients in Israel? Preliminary comment on a proposal for a bill of rights for the terminally ill. J Med Ethics 30:353–357

Barilan YM (2012) Human dignity, human rights, and responsibility. The new language of global bioethics and biolaw. MIT Press, Cambridge

Butcher T (2005) Israelis to be allowed euthanasia by machine. The Telegraph, 8 Dec. http://www.telegraph.co.uk/news/worldnews/middleeast/israel/1505018/Israelis-to-be-allowed-euthanasia-by-machine.html. Accessed 15 Jul 2015

Callahan D (1989) Killing and allowing to die. Hastings Cent Rep 19:5–6

Caplan A (2013) Brain test results don't mean Ariel Sharon will "come back," bioethicist says. NBC News, Jan 29. http://www.nbcnews.com/health/brain-test-results-dont-mean-ariel-sharon-will-come-back-1B8167234. Accessed 27 Jul 2015

Frieß M (2008) Komm süßer Tod. Europa auf dem Weg zur Euthanasie? Zur theologischen Akzeptanz von assistiertem Suizid und aktiver Sterbehilfe. Kohlhammer, Stuttgart

Ganz F, Benbenishty J, Hersch M et al (2006) The impact of regional culture on intensive care end of life decision making: an Israeli perspective from the ETHICUS study. J Med Ethics 32:196–199

Glick S (1999) The Jewish approach to living and dying. In: Pellegrino ED, Faden AI (eds) Jewish and Catholic bioethics: an ecumenical dialogue. Georgetown University Press, Washington, pp 43–53

Gottweis H, Prainsack B (2006) Emotion in political discourse: contrasting approaches to stem cell governance in the US, UK, Israel and Germany. Regen Med 1:823–829

Green RM (1999) Jewish teaching on the sanctity and quality of life. In: Pellegrino ED, Faden AI (eds) Jewish and Catholic bioethics: an ecumenical dialogue. Georgetown University Press, Washington, pp 25–42

Gross ML, Ravitsky V (2003) Israel: bioethics in a Jewish-democratic state. Camb Q Healthc Ethics 12:3247–3255

Haimes E, Williams R (2007) Sociology, ethics, and the priority of the particular: learning from a case study of genetic deliberations. Br J Sociol 58(3):457–476

Hashiloni-Dolev Y (2007) A life (un)worthy of living: reproductive genetics in Israel and Germany. Springer, Dordrecht

Howard-Snyder F (2011) Doing vs. allowing harm. In: Zalta EN (ed) The Stanford Encyclopedia of Philosophy. http://plato.stanford.edu/archives/win2011/entries/doing-allowing. Accessed 27 Jul 2015

Inbari P, Gil Y (2013) Scharons langer Weg in den Tod. Focus Online, 21 Jan. http://www.focus.de/politik/ausland/nahost/report-scharons-langer-weg-in-den-tod_aid_896832.html. Accessed 27 Jul 2015

Jacobi T, May AT, Kielstein R et al (2005) Ratgeber Patientenverfügung: vorgedacht oder selbstverfasst?. LIT Verlag, Münster

Jotkowitz AB, Glick S (2009) The Israeli terminally ill patient law of 2005. J Palliat Care 25(4): 284–288

Kongregation für die Glaubenslehre (1980) Erklärung der Kongregation für die Glaubenslehre zur Euthanasie. http://www.vatican.va/roman_curia/congregations/cfaith/documents/rc_con_cfaith_doc_19800505_euthanasia_ge.html. Accessed 27 Jul 2015

Kravitz LS (2006) "Some" reflections on Jewish tradition and the end-of-life patient. In: Hurwitz P, Picard J, Steinberg A (eds) Jewish ethics and the care of end-of-life patients: a collection of rabbinical, bioethical, philosophical and juristic opinions. KTAV, Jersey City, pp 75–98

Kröner HP (2001) „Euthanasie": von der Sterbebegleitung zum Krankenmord. In: Hucklenbroich P, Gelhaus P (eds) Tod und Sterben: medizinische Perspektiven. LIT Verlag, Münster, pp 87–103

Lepping P (2007) Ethical analysis of the new proposed mental health legislation in England and Wales. Philos Ethics Humanit Med 2:5

Nadel SH (2014) Physician-assisted suicide and the struggle for the soul of the state of Israel. The Jerusalem Post, 18 Jun. http://www.jpost.com/Opinion/Op-Ed-Contributors/Physician-assisted-suicide-and-the-struggle-for-the-soul-of-the-State-of-Israel-359823. Accessed 03 Aug 2015

National Ethics Council (2005) The advance directive: an instrument of self-determination. Opinion report. Berlin. http://www.ethikrat.org/files/Opinion_advance-directive.pdf. Accessed 27 Jul 2015

National Ethics Council (2006) Self-determination and care at the end of life. Opinion report. Berlin. http://www.ethikrat.org/files/Opinion_end-of-life_care.pdf. Accessed 27 Jul 2015

Public Committee on Matters concerning the Terminally Ill Patient (2006) The report of the public committee on matters concerning the terminally ill patient. Draft law: the terminally ill patient, 5762—2002. In: Hurwitz P, Picard J, Steinberg A (eds) Jewish ethics and the care of end-of-life patients: a collection of rabbinical, bioethical, philosophical and juristic opinions. KTAV, Jersey City, pp 207–242

Ravitsky V (2005) Timers on ventilators. BMJ 330:415–417

Ravitsky V (2006) Dying with dignity in a Jewish-democratic state. In: Hurwitz P, Picard J, Steinberg A (eds) Jewish ethics and the care of end-of-life patients: a collection of rabbinical, bioethical, philosophical and juristic opinions. KTAV, Jersey City, pp 105–118

Raz A (2004) "Important to test, important to support": attitudes toward disability rights and prenatal diagnosis among leaders of support groups for genetic disorders in Israel. Soc Sci Med 59(9):1857–1866

Schardien S (2007) Sterbehilfe als Herausforderung für die Kirchen. Güterser Verlagshaus, Gütersloh

Schicktanz S, Raz A, Shalev C (2010a) The cultural context of end-of-life ethics: a comparison of Germany and Israel. Camb Q Healthc Ethics 19(3):381–394

Schicktanz S, Raz A, Shalev C (2010b) The cultural context of patient's autonomy and doctor's duty: passive euthanasia and advance directives in Germany and Israel. Med Health Care and Philos 13(4):363–369

Schmuhl HW (2000) Nationalsozialismus als Argument im aktuellen Medizinethik-Diskurs. Eine Zwischenbilanz. In: Frewer A, Eickhoff C (eds) Euthanasie und die aktuelle Sterbehilfe-Debatte. Die historischen Hintergründe medizinischer Ethik. Campus, Frankfurt, pp 385–407

Shalev C (2000) Paternalism and autonomy in end-of-life decision-making: the Israeli normative ambivalence. Israel Yearbook on Human Rights 29:121–136

Shalev C (2009) End-of-life care in Israel—the dying patient law, 2005. Isr Law Rev 42(2): 279–305

Shalev C (2010) Reclaiming the patient's voice and spirit in dying: an insight from Israel. Bioethics 24(3):134–144

Shapira A (2006) Law and bioethics in Israel: between liberal ethical values and Jewish religious norms. J Int Bioethique 17:115–123

Simon J (2010) Ariel Sharon. Noam Braslavsky. Solo exhibition: press release. http://www.kishongallery.com/pressrelease.php?id=59. Accessed 15 Jul 2015

Steinberg A (2006) A law proposal in Israel regarding the patient at the end of life. In: Hurwitz P, Picard J, Steinberg A (eds) Jewish ethics and the care of end-of-life patients: a collection of rabbinical, bioethical, philosophical and juristic opinions. KTAV, Jersey City, pp 99–104
Steinbock B, Norcross A (eds) (1994) Killing and letting die, 2nd edn. Fordham University Press, New York
Turner L (2005) From the local to the global: bioethics and the concept of culture. J Med Philos 30:305–320
Wertz DC, Fletcher JC (eds) (1989) Ethics and human genetics: a cross cultural perspective. Springer, Berlin
Wiesing U, Jox RJ, Hessler HJ et al (2010) A new law on advance directives in Germany. J Med Ethics 36(12):779–783

Chapter 7
Lay Attitudes Towards End-of-Life Decision-Making in Germany and Israel

In this chapter we move from the medico-legal comparison, which represents national regulations and policies, to the level of personal narratives and their cultural component. National legislation as well as expert positions in Germany and Israel represent opposite regulatory approaches and bioethical debates concerning end-of-life care, as just shown. In the following, we analyze how these positions are mirrored in the attitudes of lay persons and influenced by religious views and personal experience of being affected. We qualitatively analyze eight focus group discussions in Germany and Israel, where lay persons (religious, secular, affected, and not affected) were asked to discuss similar scenarios involving end of life decisions.

Overall, in both countries, respect for patient autonomy and wishes to die with dignity found broad consent. Lay people argued in favor of the acceptance of wishes put down in an advance directive. Based on respect for autonomy, lay persons in non-religious groups in both countries argued for possibilities of active euthanasia in severe cases—but at the same time cautioned against possible misuse. National contrast was apparent in the moral reasoning of lay respondents concerning the distinction of withholding and withdrawing treatment. Especially modern religious lay persons in Israel argued strongly against allowing the withdrawal of treatment based on a patient's wish, referring to halakhic tradition. We conclude by discussing the emergent notion of shared responsibility and views of professional responsibility, connecting them with relevant cultural themes such as religion and national culture.

Evidently, personal narratives of a good death are influenced by cultural scripts that may change from country to country (Long 2004). In post-industrial societies, such as Germany and Israel, there are expected trajectories of dying which differ from those of earlier eras and less industrialized nations. For example, a person in industrialized countries is more likely to die in old age than in childhood, and death is more likely to occur following chronic illness than from infectious disease. Death in such countries most often takes place in hospitals, attended by "experts" and medical machines (Seale 1998, 2000). Care of the dying has been largely institutionalized and commodified; at the same time, it is widely assumed that there is now an element of

A.E. Raz and S. Schicktanz, *Comparative Empirical Bioethics: Dilemmas of Genetic Testing and Euthanasia in Israel and Germany*, SpringerBriefs in Ethics, DOI 10.1007/978-3-319-32733-4_7

choice in how one dies. While this phenomenon is described widely for many countries and seen as a grand narrative of ageing and dying in late modernity (Imhof 1981; Illich 1975; Ariès 1981), differences in closely related cultural contexts were overseen for a long time.

A particular sensitive issue affecting deliberations in Germany and Israel is the Holocaust and the role of Nazi doctors in non-voluntary euthanasia. The experience of the political instrumentalization of medical experts is often raised in the German discourse. It is used as historical evidence for sustaining slippery-slope arguments and opposing any form of doctor-assisted suicide or active euthanasia, a practice now common in Belgium, Switzerland, and the Netherlands. Interestingly, the attitudes of Israeli Holocaust survivors concerning euthanasia (although playing a minor role, if any, in the emerging legal policy) were found to be opposed to those of German professionals (Leichtentritt and Rettig 1999). Israeli Holocaust survivors argued that profound differences exist between Nazi practices of non-voluntary euthanasia and physician-assisted dying based upon a patient's wish. Therefore, they cautioned against comparisons between the Holocaust and other practices. Cultural factors, such as religion and the lessons of the Holocaust, influence not only the process of legislation but also the attitudes of health professionals (Pelleg and Leichtentritt 2009; Sprung et al. 2007; Wenger and Carmel 2004) and the role played by patient support groups (Raz et al. 2014).

Recent studies have highlighted the importance of cultural factors for end-of-life decisions (Gysels et al. 2012), especially for maintaining an interface between health professionals and family members for joint decision making in the sensitive context of end-of-life care and euthanasia. Taking patients' and families' cultural background and position between individualism and collectivism into account is recommended for structuring decision situations in accordance with a patient's wishes (Searight and Gafford 2005). The bioethical issue of "shared decision-making" in end-of-life has emerged as reflecting this interactive locus of decision making and is taken up in both frameworks. In Israel, patient involvement and autonomy in decisions is strengthened by legal framework, but current practice shows a relative lack of shared decision-making also in the context of end-of-life (Miron-Shatz et al. 2011). For Germany, Härter and co-authors show that, besides a variety of activities and training programs for shared decision-making, work needs to be done to establish shared decision processes as routine also in the context of end-of-life decisions (Härter et al. 2011).

We examine the attitudes and arguments of lay people toward end-of-life decisions regarding the distinction between withdrawal/withholding treatment[1] and

[1]We differentiate between withholding of treatment (not administering it when indicated) and withdrawing treatment (stopping already started treatment; e.g., stopping artificial ventilation), which are summarized under the German term "passive Sterbehilfe" (passive euthanasia) as two ways of letting a patient die. Furthermore, we discuss physician-assisted suicide and euthanasia as two acts that aim at deliberately ending a person's life. In the first case the act is done by the patient himself (e.g., applying a deadly dose of treatment); in the second case the application is administered by a physician.

the application of so called advance directives as well as forms of euthanasia. Euthanasia is often differentiated into passive and active euthanasia. Passive euthanasia is about withholding treatment (e.g., choosing not to connect a dying person to life-support machines). Active euthanasia is defined as the application of a lethal dose of a substance to terminate life of a patient. Further differences are made, whether the application is done by the doctor or whether doctors provide only the substance, but the final act is controlled by the person herself (so called physician assisted suicide). Such accepted terminologies are seen as important if we are interested in comparing different attitudes and practice (see Van der Heide et al. 2003; Sprung et al. 2007). However, as our study will show, these academic definitions do not always match the understandings and interpretations of lay persons.

Participants for the focus groups were chosen in a way that reflects different backgrounds with regard to "being affected." Being affected was defined as having cared for a terminally ill person in contrast to persons who had had no such experience (see Schicktanz et al. 2008).[2] Two groups in each country were conducted with affected persons, and two groups with non-affected persons. In order to be able to assess the relevance of religion and religious arguments, one group of non-affected persons was composed of participants with a religious background and contrasted by a group with secular participants (based on the self-ascription of participants). The participants of the "modern religious" groups were recruited from Christian parishes in Germany and Jewish communities in Israel. We were especially interested in how far modern religious participants referred to arguments used by religious leaders in the public debate on end-of-life legislation. The participants of the focus group discussions were recruited by information sheets, flyers and posters in public places.

Focus groups had between 5 to 9 participants with 59 participants in total, 25 male (42 %) and 34 female (58 %). Participants' age was between 20 and 80 years, with different age groups represented in all focus groups. Participants had different educational backgrounds, but self-recruitment resulted in a slightly higher share of people with higher levels of education. In all groups the same scenarios and questions were used. After the presentation of end-of-life scenarios, participants were asked to discuss arguments for or against withholding or withdrawing a suggested treatment. The first scenario was about an elderly woman (89 years old) who is unconscious and hospitalized for some time. The relatives present an advance directive which instructs not to use life-sustaining treatment, such as artificial support of heart- and lung-function by a machine (withholding treatment). The attending physicians are unsure if they should follow it in case of a heart or lung failure. In the second scenario the same person was already connected to a respirator after having been admitted to hospital because of a heart attack. The attending doctors are unsure if they should disconnect the artificial respirator

[2]We did not ask terminally ill patients themselves to participate since discussing end-of-life decisions might have been too tiring, stressful and disturbing for them.

(withdrawing treatment). Furthermore, participants were confronted with a scenario of a patient who suffers from untreatable pain and approaches the doctor to help end the futile suffering.

7.1 Patient Autonomy as the Leading Principle for Withholding End-of-Life Medical Treatment

Participants in all of the groups of both countries agreed that treatment should only be started if it is in accordance with the patient's will. If a patient has, as in the first scenario, an advance directive that does not allow life-sustaining treatment, artificial ventilation should not be started. All of the participants judged unanimously that it is the responsibility of the patient's doctor as well as her family to adhere to her wish:

> Yes, do it! Because doctors should respect the request of a dying patient, who knows what is best for her. (Israeli female, affected group 1)

The advance directive was seen as an expression of one's will which should be as respected as direct communication with a doctor.

> Physicians have to implement the written instructions because it is the same as asking the woman herself, she has written down in advance whether she wants to get help from machinery like CPR (cardio-pulmonary respirator) or anything like that, when she still [has] the ability to decide about her own life. She asks not to go against the natural process of dying and not to interfere deliberately, intervention is not natural. It is the duty of doctors to implement her request. (Israeli female, affected group 2)

In order to strengthen their views on respecting the advance directive and patient autonomy, participants referred to the patient's dignity. Dying with dignity was described as an integral part of the dignity of life and as implying the ability to act autonomously and that others ought to respect one's wishes.

Discussions of "dying with dignity" were connected to images of old age in all groups with slight national differences. In the Israeli groups, old age was seen as a good reason for respecting a person's decision. In the German groups, old age was regarded as an indicator of a life fully lived and as calling for reflection on the value of successfully finishing a life course. Discussions of the desire for the withholding of treatment were embedded with images of a good death, and participants claimed that this desire should be respected, as the following quote shows.

> And I think, at a very old age, sometimes it is somehow … One has likewise to see that one dies in dignity. If someone says: "I want that with 85 years of age my life is concluded. I had a beautiful life." And then one should, the way I see it respect this wish. (German female, modern religious, non-affected group)

However, the German groups also discussed the notion of a life fully lived critically as something subjective that can be assessed only by the patient herself and not from a third party's perspective. Otherwise, there might be the risk that elderly people are seen generally as dispensable and become victimized by age rationing:

> Stop! Age does not matter at all. Age plays here … Then we could simply put our old people down … (German female, secular, non-affected group)

A few participants in both countries wanted to restrict the scope of self-determination for withholding treatment. They voted for allowing the withholding of treatment only in cases where there is no hope of a cure. So, the patient's wish not to receive treatment should be followed only when the dying process has already started. For this minority, the autonomy of the patient remained an important principle, but the impossibility of recovery was introduced as a restricting criterion.

> But if it is simply the case that she remains confined to bed for a week and the doctors say, "Well, let's see, there exists the possibility that she will become independent again and can live somewhere in a retirement home or her private apartment," then it is certainly clear that one must perhaps do much more. But if the only concern is life support, then it is the age on the one hand, and, of course, dignity on the other hand. (German male, affected group 2)
>
> I agree that the question here is, if this is final? For the patient I mean. If the situation is final and she was asking for it, then they [doctors] should comply. (Israeli male, modern religious, non-affected group)

7.2 Withdrawing Treatment: Between Patient Autonomy and Doctors' Duty "Not to Harm"

While discussions of withholding treatment were rather uncontroversial the idea of withdrawing a treatment, like artificial ventilation, that has already been started caused more debate. The fact that withdrawing treatment necessarily involves the active intervention of doctors received special attention in the reflections of participants. When discussing the second scenario, in which artificial ventilation was already in use and doctors learned only later about the advance directive and the patient's decision against life-sustaining treatment, more diverse positions evolved in both countries.

In the Israeli focus groups, there was a general tendency against withdrawing treatment. The main reason given was that withdrawal requires the active intervention of doctors and that was seen as morally different from passively withholding treatment.

> There is no justification to disconnect the patient if the connection to the machine is already made and is planned as a short-range intervention. (Israeli male, affected group 1)

Participants did not regard the patient as dying but connected her death to the action of the doctor. Therefore, withdrawing treatment was not considered ethically equivalent to withholding it:

> This question is more difficult because it is like killing her. (Israeli female, secular, non-affected group)

Participants in the modern religious group in Israel especially emphasized that withdrawing treatment is more problematic than withholding it. They referred to Judaism in general and Jewish law in particular.

> The law in *Shulchan Aruch* states that you may not even move a pillow from underneath the head of a dying person in order to hasten his death. (Israeli male, modern religious, non-affected group)

Among the German groups, positions were not as clear-cut. Although there was a general tendency to see withholding and withdrawing treatment as similar moral predicaments, as both lead to the patient's death, the second scenario induced more controversy. Participants in the modern religious group, especially, saw a morally relevant difference between withdrawing and withholding treatment.

> Well, I see a great difference there [between withholding and withdrawing]. […] here we have the concrete case where an apparatus is connected, and – that when it is disconnected – the patient will die. That does not have to be the case. There is no guarantee that the person will indeed not continue breathing. But as long as the machine is connected, the person will breathe. (German male, modern religious, non-affected group)

Furthermore, members of the German focus groups discussed the individual situation and, especially, the chances of recovery from it intensively.

> I would say that the question is: "Who is the patient? What are the chances of recovery?" Not the age is crucial, but what the doctors think. One has to somewhat rely on the doctors. (German female, secular, non-affected group)

When there was no chance of recovery, there were strong votes in both countries—though not an overall consensus—in favor of respect for the patient's self-determination and the withdrawal of treatment. The timing of the decision played a crucial role here:

> These interventions have been made, nothing can be done. But if it lasts for a long time – I estimate around two weeks – or three, I think you must disconnect according to the request of the patient. (Israeli male, affected group 1)

In both the Israeli and German groups, the process of decision-making and the distinction between long-term and short-term interventions were raised in attempts to resolve such a dilemma. There was overall agreement that in an emergency any treatment to help the patient should be given immediately. Participants in all of the groups, however, were concerned that such cases can lead to the dilemma of withdrawing treatment later on if the emergency intervention does not restore the patient. Preferably, such situations should be avoided by prior extensive communication between patients, relatives and doctors about treatment plans and advance directives. Ideally, the question of withdrawing treatment will not arise when treatment is started only if it is effective and accords with the patient's wish. Given the complex chain of decisions that end-of-life care requires, participants in all groups agreed in rejecting automatic emergency treatment. There was also strong support in both countries for respecting patient autonomy even in cases of withdrawing treatment.

> On my part, I would say, that first of all, the apparatus should be turned off because this is what the patient wanted. (German male, modern religious non-affected group)

However, even when participants argued in favor of withdrawing treatment, they raised concerns about the appropriateness of requesting this of doctors. Participants in both countries who followed this line of argument expressed empathy with the doctor who might find withdrawing treatment morally problematic.

> If the situation deteriorates and there are no chances for recovery, the advance directives should come into effect, within the framework of the Israeli law. However, I cannot imagine the possibility that someone takes responsibility for another's life, I mean actually disconnecting the patient from the life-support machines. (Israeli female, modern religious, non-affected group)

Therefore, the law should protect doctors.

> It has to be clearly regulated that if it is on the paper there are no legal consequences for the person who pulled the plug. (German female, modern religious, non-affected group)

While the law cannot free doctors from the responsibility for withdrawing treatment, it should at least provide a clear framework in which to make decisions. Participants were not always familiar with their countries' current legal framework.

7.3 Physician-Assisted Suicide and Euthanasia: Unsolvable Dilemmas and the Advice to Be Cautious

The third scenario involved an everyday language plea for assisted suicide or euthanasia. Participants in all groups found it important to express their acknowledgment of the difficulty of the ensuing moral dilemma. Overall, they were very sympathetic with patients who wanted to end their suffering. Affected and secular laypeople in both countries openly discussed ways of relieving the patient's pain. The reasons given dealt primarily with the individual case presented, and participants found it difficult to provide clear-cut suggestions for regulations applicable to the case.

> My opinion on that is a bit paradoxical. In this concrete example, I would be in favor of the doctor following it. But I would likewise be for if, given that it is illegal, that [he] is prosecuted for it. Naturally, I also don't want a legal vacuum in this situation, for one thing, and secondly, I don't want it to be legalized in Germany. For the simple reason that this leads to a slippery slope, as one says in English; where one would quickly start drifting off. (German male, affected group 2)

> This moral dilemma is very heavy. I'm all in favor of helping – not to continue agony for days and months. I cannot morally oblige the doctors to give him something to hasten the death, and for some reason I cannot explain, something inside my feeling, I think that refraining from treatment that preserves the life is preferred in this case. To do an act of active killing is morally difficult for me, though both options are similar in terms of the end results. (Israeli male, affected group 1)

The heterogeneous suggestions ranged from options within the current legal framework, like additional palliative care, palliative sedation, and stopping treatment, to physician-assisted suicide including travel to countries that allow it.

Participants who had experienced a close relative's suffering and death recounted stories that were similar to the scenario, as in the following example:

> And there were moments of weakness where he [the father] said: "Just take a stick and kill me." Then as a relative you are standing at the bedside and say [to yourself]: "What does he tell now?" But he is completely lucid. But not in a way that [hesitates] all is still intact. Simply, that he could not take the suffering anymore. (German male, affected group 1)

However, the experience didn't make it any easier for participants to come up with solutions. Answers were often openly paradoxical. On the one hand, participants saw that the relief of suffering was an important aim; on the other hand, they raised serious concerns about legalizing physician-assisted suicide or even euthanasia. They were especially concerned about possible abuses of liberal euthanasia regulations:

> But if I find myself in the position where the person in front of me tells me faithfully: "That is different for me, and I really want that," then I am not the one who has to make the decision. Insofar, I think, I would be ready to give her these drugs. Precisely because the decision does not lie with me. Whereby – since it actually concerns euthanasia – I have to remark that I nevertheless find that is should be forbidden since the danger of abuse is far too high. (German male, modern religious, non-affected group)

All of the groups discussed different types of social pressure, from family members as well as the broader society, coercing patients to hasten their death against their will as a serious risk; the main concern was that relatives might want to get rid of the elderly for financial reasons:

> Unfortunately, in some cases we heard that family members take advantage of an incurable illness of a family member to get money, and their decision to disconnect him from life support, for example, does not seek to benefit the patient." (Israeli male, affected group 1)
>
> It is also a problem if, for example, in a nursing home relatives would say: "We don't want that anymore." – Do you now act by thinking of the resident, who perhaps does not have an advance directive, or are there some among them who would like to save on the 1500 euros nursing home costs per month or get the inheritance. (German female, affected group 2)

To contrast the more individual case-centered perspective, participants were also asked to comment on the legalizing of physician-assisted suicide and euthanasia. Overall, participants in both countries, except for the modern religious Jewish groups, were in favor of changing the current law and allowing these actions. They advocated clear regulations and effective controls to prevent abuse but stressed that new permissive legislation should apply only to terminally ill patients with unalleviable suffering.

> … yes, not just one and very quickly, but two doctors and perhaps yet another person. Well – caution! (German female, affected group 2)

In contrast to the other groups, modern religious participants from Israel unanimously opposed assisted suicide and any liberal regulation of euthanasia.

> Yeah, that is a problem. I don't think that Judaism teaches in favor of prolonging the dying process. But you certainly cannot hasten death. (Israeli male, modern religious non-affected group)

They explained their straightforward position in terms of religious values, such as the sanctity of life.

> I think the doctors should not help him with this because God gave him his life and is the only one who may take it away. (Israeli female, modern religious, non-affected group)

7.4 Who Should Decide? Shared Decision-Making as a Challenge

There was general agreement in all of the groups that an advance directive does not automatically lead to a clear decision with which everyone can agree. Advance directives need to be applied, and this can result in new ethical dilemmas. There was an overall consensus that decisions on behalf of patients are difficult to make and that responsibility for them is a burden that causes anxiety.

Participants were asked which additional procedures should be in place in case of conflict over the application of an advance directive. Overall, opinion was equally divided between giving the final authority to the family or the doctor. Most importantly, though, the decision should reflect the patient's will and take the advance directive as a guideline. While the German groups tended to trust family members more than doctors, many Israeli participants strongly advocated for doctors, pointing out that they are experts and have the most experience with end-of-life decisions. Arguments in favor of doctors as final decision-makers were that they have a better clinical understanding of the case, can better assess what is in the patient's interest, and are not emotionally involved.

> The doctors are the most professional and they really know the situation – if there is a chance, what are the chances, these are things that the family or the court cannot know without the doctors. (Israeli male, modern religious, non-affected group)

Since doctors are bound by the ethos of their profession to cure, most German participants favored giving family members the authority to make decisions. Their arguments were that they know the patient best and therefore understand what he would want in a specific situation. As two participants put it:

> The relatives.— Of course, always on the basis of the advance directive. (German female, affected group 2)

> [R]elatives should have the final say, since these are the persons who know the patient best and for whom the patient's life means most. (German female, modern religious, non-affected group)

Some participants in both Germany and Israel who had had no experience with making end-of-life decisions mentioned the problem that family members are emotionally involved and would find it difficult to decide for their loved ones.

> Well, the difficulty that I see is that if the responsibility to decide lies with the relatives, due to the closeness of the relationship, anxieties and the inability to say goodbye of relatives will perhaps be an important issue. (German female, modern religious, non-affected group)
>
> The relatives are very involved emotionally and it's also not very fair to put them in this situation where they should choose. The doctors can make a rational decision, according to the specific therapeutic risks and benefits, but for that – so they can really make a balanced decision and not be afraid that somebody would sue them, the law should be on their side. (Israeli male, secular, non-affected group)

Interestingly, participants in the affected groups had no such doubts, and this supported their desire to be involved in decision-making regardless of the burden. There was an overall agreement that a shared decision process, in which different perspectives are heard and a solution is found together, is best. Related to this attitude is the belief that only those involved in the situation should participate in decisions. Thus, the court as impartial party was seen as a last resort which should have no authority of decision making but which could be appealed to in cases of disagreement.

> In case of disagreement between the doctors and the relatives – the court should decide. (Israeli male, modern religious, non-affected group)

7.5 Formal or Informal Advance Directives?

Throughout all of the discussions, participants talked about the ambiguity of an advance directive that results when the patient's wishes are not expressed clearly. Participants in all groups agreed that advance directives should meet formal requirements, which one participant summarized as follows:

> They must fully accept the request of the woman, but only if she wrote the directives in clear and sound mind and with full understanding of the issue, in case she has no medical background regarding the implications of her request. (Israeli female, secular non-affected group)

Participants were also asked about the advantages and disadvantages of formal and informal advance directives. Among the Israeli groups, there was agreement that the formal form mandated in Israel is preferable to the informal form. They argued that a precise advance directive based on the mandated form is the most helpful sort in decision-making.

> The benefits of the formal form – it is very clear. Not too many gray areas. (Israeli male, secular, non-affected group)

The form's complexity (and the legal regulations) forces people to seek out advice and inform themselves about all of the options and their consequences. Patients' thorough understanding of the matter due to that was seen as a safeguard.

German participants were less clear about the form that advance directives should take. While many also argued that a precise advance directive has the advantage of clearly communicating the patient's wishes to doctors, some supported the informal form. According to them, an advance directive that provides information about a patient's personality from their own point of view rather than just ticking boxes about which treatments are declined under which medical conditions provides a better guarantee that treatment accords with the patient's wishes.

> Well, I rather would like to write down my own thoughts. What I think of life; what my values are and how I would imagine it to be then. (German female, secular non-affected group)

Furthermore, German participants thought that the large number of templates for advance directives is more confusing than helpful.

7.6 Conclusions

End-of-life decisions and the problems related to them are intensively discussed in societies with advanced medical systems. Thus, in the beginning of the 21st century dying is no longer the taboo that, in the 1980s, the social historian Philippe Ariès claimed it to be (Ariès 1981). Instead, the public discussion of difficult cases and of the process of shaping a legal framework for advance directives has transformed the previous silence over dying in a hospital into a public awareness of the difficulties and dilemmas involved. The results presented here provide insights into the complex ways that cultural factors and experience with the end of life influence how laypeople perceive end-of-life decisions.

While we found that being affected is an important influencing factor in many fields of medicine (see previous chapters,) where the power of lived experience overrules particular cultural perspectives (Kleinman 1999), the case of end of life issues might differ.

Our study shows no such strong differences of attitude between affected and non-affected persons in the context of end-of-life issues. There was overall consensus, cutting across affected and non-affected groups, regarding general issues such as supporting the development of institutions for palliative care, respecting patient autonomy, and defending patients' decisions to limit treatment. Participants also agreed that it is necessary to proceed with caution in end-of-life situations and strive for shared decision-making involving all stakeholders. One possible explanation for this is that the media discourse on demographic change and end-of-life dilemmas has increasingly influenced the general public in many Western, industrialized countries. In contrast to other medical settings involving life-planning, such as genetic testing or organ transplantation, advance directives are relevant to everyone, and, consequently, the issues connected to end-of-life decisions are well-known to the public. Thus, participants stressed the importance of autonomy as much as experts do (Gedge et al. 2007). The most significant argument reflecting the unifying perspective of being affected was the support for euthanasia in extreme cases where the risk of losing one's autonomy and dignity overruled the perceived responsibility of society to regulate end-of-life care. The unanimous claim for more self-determination and professional respect for patients' (and citizens') autonomy is a result of the shared belief that everyone is capable of her own end-of-life decisions.

Shared deliberation about palliative and curative therapy options and underlying should be the basis of good decision-making. In shared deliberation, as the participants in our focus groups regarded it, advance directives, as the voice of the patient himself, are essential because family members or third parties do not always properly take their position into account. This is in contrast to the opinions of some experts who defend the power of attorney as the more powerful instrument for guaranteeing patient autonomy (Fagerlin and Schneider 2004). In the light of their experiences of family struggles, our participants criticized the claim that a power of attorney is an adequate substitute for an advance directive.

Our study indicates the need for critical reflection of the classical model of doctor-patient communication (Emanuel 2008; Hanson 2008). Shared decision-making in different fields of medicine has already been thoroughly discussed, and various studies have shown its positive effect on decision outcomes (see Loh et al. 2007; Joosten et al. 2008). However, the participants in our study conceptualized this complex communication in terms of shared responsibility. They extend the classical model by explicitly integrating normative aspects of the process into decision-making. As they understand it, the decision-making process should be a complex deliberation among moral agents about questions of dying with dignity and the quality of life in which doctors are stakeholders with a professional ethos. They regarded the ethos of doctor to cure and to save life as the source of their expertise and authority, but that authority, they also thought, needs to be counterbalanced in the context of the end of life to prevent the prolonging of life by any means. Respondents in the German focus groups, especially, stressed this. They claimed that medical authorities are too biased (in favor of life) to make objective end-of-life decisions but that they nevertheless constitute a morally important perspective, though one that should be complemented by the perspective of family members. Those lay perspectives have equal standing with medical ones when it comes to determining what constitutes living and dying with dignity.

The concept of shared responsibility can be translated into the role assigned to the family in such a decision process. Differences with regard to the cultural scripts of families thereby became apparent. While laypeople in both countries generally expressed a desire that relatives be involved in the decision-making process, German participants were concerned about the social challenge for those patients who have no family or are not on good terms with their family. In Israel, a more traditional image of the family prevailed in which family, as an omnipotent source of support, is always there (see Birenbaum-Carmeli 2010; Hashiloni-Dolev and Shkedi 2007).

Our research design also gave us a more nuanced insight into the interdependency of culture and religion. While the difference between the Christian, German and Jewish, Israeli groups was relatively small, the difference between religious and secular groups was more apparent. Respondents in the religious groups in both countries tended to regard the difference between withholding and (actively) withdrawing treatment as more ethically problematic than participants from the secular groups. Our findings support some of the results of the quantitative ETHICUS study (Sprung et al. 2007), which emphasized the differences between religious and secular individuals. The ETHICUS study found strong effects of enculturation for Protestant/Catholic doctors depending on whether they were working in Southern, Central, or Northern Europe. These results are similar to our findings that religious denomination needs to be studied in its enculturated form and local context. Differences in, e.g., legal contexts, cannot be explained solely by the influence of different religious backgrounds or of different religious leaders, despite their prominent roles in national debates. The group of non-religious persons that is not represented in the same way like religious groups in the public debate and their negotiation with religious position needs to be further investigated.

Interestingly, drawing a line between actively and passively intervening at the end of life remains a challenging question. In all of the groups in both countries, participants raised the problem that doctors hesitate to withdraw treatment, e.g., switch a machine off, because it feels like an active intervention that leads to a patient's death. Surveys and qualitative studies of German doctors have shown that doctors, despite professional training, still differentiate between withdrawing and withholding treatment. Therefore, they tend to misjudge which forms of refraining from treatment are in accordance with German law and which are prohibited (Borasio et al. 2004; Van Oorschot et al. 2005; Beck et al. 2009). In theory, most ethicists and lawyers favor treating both withdrawing and withholding treatment as equivalent decisions to limit therapy. Distinguishing active and passive intervention does not necessarily lead to different judgments about the kinds of end-of-life actions that are ethically justifiable. Our participants voted qualitatively for allowing both by giving priority to the patient's desire. As an input for ethical theory, we suggest that acknowledging the moral psychological challenge of withdrawing treatment should be further considered. Instead of insisting on a formalistic theoretical approach, one informed mainly by the analytical and consequentialist tradition, which equates both actions, addressing this difference from a moral psychological point can help to improve real-life deliberations.

References

Ariès P (1981) The hour of our death. Knopf, New York

Beck S, Van de Loo A, Reiter-Theil S (2009) A "little bit illegal"? Withholding and withdrawing of mechanical ventilation in the eyes of German intensive care physicians. Med Health Care Philos 11(1):7–16

Birenbaum-Carmeli D (2010) Genetic relatedness and family formation in Israel: lay perceptions in the light of state policy. New Genet Soc 29(1):73–85

Borasio GD, Weltermann B, Voltz R et al (2004) Einstellung zur Patientenbetreuung in der letzten Lebensphase: eine Umfrage bei neurologischen Chefärzten. Nervenarzt 75(12):1187–1193
Emanuel L (2008) Advance directives. Annu Rev Med 59:187–198
Fagerlin A, Schneider CE (2004) Enough: the failure of the living will. Hastings Cent Rep 34:30–42
Gedge E, Giacomini M, Cook D (2007) Withholding and withdrawing life support in critical care settings: ethical issues concerning consent. J Med Ethics 33:215–218
Gysels M, Evans N, Meñaca A et al (2012) Culture and end of life care: a scoping exercise in seven European countries. PLoS ONE 7(4):e34188
Hanson JL (2008) Shared decision making: have we missed the obvious? Arch Intern Med 168(13):1368–1370
Härter M, Müller H, Dirmaier J et al (2011) Patient participation and shared decision making in Germany—history, agents and current transfer to practice. Z Evid Fortbild Qual 105(4):263–270
Hashiloni-Dolev Y, Shkedi S (2007) New reproductive technologies and family ethics: pre-implantation genetic diagnosis for sibling donor in Israel and Germany. Soc Sci Med 65(10):2081–2092
Illich I (1975) Medical nemesis: the expropriation of health. Calder & Boyars, London
Imhof AE (1981) Die gewonnenen Jahre. Von der Zunahme unserer Lebensspanne seit dreihundert Jahren oder von der Notwendigkeit einer neuen Einstellung zu Leben und Sterben. Ein historischer Essay. Beck, München
Joosten EA, DeFuentes-Merillas L, de Weert GH et al (2008) Systematic review of the effects of shared decision-making on patient satisfaction, treatment adherence and health status. Psychother Psychosom 77:219–226
Kleinman A (1999) Moral experience and ethical reflection: can ethnography reconcile them? A quandary for "the new bioethics". Daedalus 128(4):69–97
Leichtentritt RD, Rettig K (1999) Meanings and attitudes toward end-of-life preferences in Israel. Death Stud 23(4):323–358
Loh A, Simon D, Wills CE et al (2007) The effects of a shared decision-making intervention in primary care of depression: a cluster-randomized controlled trial. Patient Educ Couns 67(3):324–332
Long S (2004) Cultural scripts for a good death in Japan and the United States: similarities and differences. Soc Sci Med 58:913–928
Miron-Shatz T, Golan O, Brezis M et al (2011) The status of shared decision making and citizen participation in Israeli medicine. Z Evid Fortbild Qual 105(4):271–276
Pelleg G, Leichtentritt RD (2009) Spiritual beliefs among Israeli nurses and social workers: a comparison based on their involvement with the dying. Omega 59(3):239–252
Raz A, Jordan I, Schicktanz S (2014) Exploring the positions of German and Israeli patient organizations in the bioethical context of end-of-life policies. Health Care Anal 22(2):143–159
Schicktanz S, Schweda M, Franzen M (2008) "In a completely different light"? The role of being affected for epistemic perspectives and moral attitudes of patients, relatives and lay people. Med Health Care Philos 11(1):57–72
Seale C (1998) Constructing death: the sociology of dying and bereavement. Cambridge University Press, Cambridge
Seale C (2000) Changing patterns of death and dying. Soc Sci Med 51:917–930
Searight HR, Gafford J (2005) The cultural diversity at the end of life: issues and guidelines for family physicians. Am Fam Physician 71(3):515–522
Sprung CL, Maia P, Bulow HH et al (2007) The importance of religious affiliation and culture on end-of-life decisions in European intensive care units. J Intensive Care Med 33(10):1732–1739
Van der Heide A, Deliens L, Faisst K et al (2003) on behalf of the EURELD consortium. End-of-life decision-making in six European countries: descriptive study. Lancet 362(9381): 345–350
Van Oorschot B, Lipp V, Tietze A et al (2005) Einstellung zur Sterbehilfe und zu Patientenverfügungen. Ergebnisse einer Befragung von 727 Ärzten. Dtsch Med Wochenschr 130(6):261–265
Wenger NS, Carmel S (2004) Physicians' religiosity and end-of-life care attitudes and behaviors. Mt Sinai J Med 71(5):335–343

Chapter 8
Risks and Responsibilities: Making Plans for Life and Death

In this chapter we would like to bring the concept and perceptions of risk further into the discussion. Responsibility and risk are two major elements of health promotion whose linkage has been acknowledged, for example, in the context of smoking, exercise, low-cholesterol and low-fat diets, and other lifestyle changes that increasingly are part of our "new health morality" (Conrad 1994). However, as new medical technologies bring about new (likely and uncertain) risks, which are not only technical or medical risks, but also social risks, the social expectation to "act responsibly" can also take new forms and interpretations. For example, genetic testing informs a person about her individual risk in statistical terms to suffer from colon cancer, but there is also a risk that this information is wrong. Statisticians call this "second-order risk." Moreover, there is a risk that the knowledge of this statistical information will lead to social implications such as restrictions for employment or higher fees for health insurances. One can say, as the latter is not yet put into statistical terms, this is not a risk but a danger or—as philosophers would suggest—it is a slippery slope argument for future implications. However, all three dimensions share the idea of an anticipated but not 100 % certain future outcome, and this outcome is rated negatively.

The meaning of "risk" is therefore dependent on the medical context we focus on: the definition and perception of risk is extended to embrace new ways of conceptualizing uncertainty (as in the case of predictive genetic testing) or potential loss of autonomy (as in the context of end-of-life care and advance directives). In the following, we want to compare the effect of culture and susceptibility on lay people's attitudes towards responsibility and risk in the context of two morally loaded medical contexts: predictive genetic testing (and its relation to eugenics); and end-of-life care (with relation to euthanasia).

The findings are organized in this chapter according to the three main themes of responsibility that dominated the discussion across all groups. They characterize three different types of relationships: (a) self-responsibility as an expression of a moral self-relationship and understanding of oneself as a moral actor, (b) the individual responsibility towards family members based on the basic experience of

A.E. Raz and S. Schicktanz, *Comparative Empirical Bioethics: Dilemmas of Genetic Testing and Euthanasia in Israel and Germany*, SpringerBriefs in Ethics, DOI 10.1007/978-3-319-32733-4_8

being a social entity related to at least a few but very close fellow human beings, and (c) the responsibility of society towards individual members as a modern understanding of what is the basic role of the state or society towards its individual members. These three themes are matched with a corresponding perception of risk.

8.1 Self-responsibility and Individual Risk

A common argument across the groups in favor of testing for genetic risk factors was the availability of preventive or curative strategies, for example, in the context of colon cancer, as already discussed in Chap. 4. This argument stresses that because you can perform a surgery to cure the disease when detected early, it is better to perform the test as early as possible in order to undergo the operation in time. Others mentioned in this context that if one knows that there exists a risk one will be examined more frequently. However, while the majority of Israeli respondents voiced this argument, only a minority of German respondents agreed to it. Many German respondents argued against taking the test in the colon cancer scenario because of the problem of probabilistic knowledge that creates anxiety, and implies the danger of discrimination. This argument stresses that we should live in a society where we allow the existence of risks. The largest support of taking the test for colon cancer, as already explained in Chap. 4, was expressed in the affected German group. Indeed, this argument was supported first and foremost through personal experience of cancer and its risk management.

When the risk was still there but without options for prevention (the Huntington's disease scenario), individual responsibility for taking the test was seen as less binding. The majority of the respondents across all groups, in Israel and Germany, were thus against taking the test in the context of Huntington's disease because even though it provided certainty regarding the risk, this risk knowledge cannot match with a potential benefit but might rather create social and psychological risks. Indeed, some respondents argued that self-responsibility meant refraining from taking the Huntington's disease genetic test, which could entail further damage in the form of anxiety.

It seemed important for our respondents to discriminate "self-responsibility" based on the duty to stay healthy/to promote well-being, and the virtue of self-care. The latter is a less strict and rigorous moral principle. It can be seen as moral orientation by which we consider striving for a wise and reflective handling of one's own body-soul relationship as part of a good life. The common claim that genetic responsibilization results rather in an unreflective embracing of a duty to be tested (Arribas-Ayllon et al. 2008) might be true for some people; however, these findings highlight that there can be many more nuanced and diversified interpretations.

In the context of end-of-life decision-making such as preparing advance directives, some authors have also criticized an increasing form of responsibilization and self-government with the wish of "letting someone die" or with "assisted suicide" (see, e.g., Rose and Rabinow 2006). However, we cannot really empirically confirm this finding. The common, prevailing moral attitude consisted rather of the claim that whatever a person wants concerning his or her death needs to be respected by others. This holds true as long as it does not impede strong moral feelings on others (as the request of euthanasia might impose on others, such as doctors). Notably, the general awareness and low impact of being affected we found here can tentatively be interpreted as a finding that modern citizens have nowadays internalized the need to consider end-of-life issues in advance. This differs from other medical issues often only addressed and discussed by patients and their close surroundings. It would be worth examining also in the future in more detail whether this can be called a trend turning away from the practice of denying death. The "denial hypothesis" was brought up in the second half of the 20th century (Becker 1973; Ariès 1981; Zimmermann and Rodin 2004). Others are currently speaking of the "garrulity" of death (Nassehi and Saake 2005) as death and dying are major themes in late modernity and its pop culture.

8.2 Family Responsibility and Shared Risk

Family responsibility encompasses both individual responsibility for other family members or the family as a group as well as the collective responsibility for an individual family member or the family as a group. Why put these different cases (individual for individual; individual for collective; collective for individual; and collective for collective) into one category one could ask. And indeed, there are cases where attention must be given to this difference within the conglomerate. However, for analytical reasons, we found it more appropriate to merge these different dimensions into one scheme as the actor's identity is mainly shaped by the internalized social role of being a family member. This social role does conflate individual and collective perspectives.

Arguments related to family responsibility were immediately apparent in both bioethical cases of genetics and end of life. Others have also shown their high socio-empirical relevance for discussing organ donation (Wöhlke 2014, 2015); nutrition and health (Turoldo 2009) and, most obviously, repromedicine (Beier et al. 2015; Wiesemann 2010). Our respondents extended the discussion of responsibility in the context of genetic testing by relating it to what they perceived as responsibility towards one's family members. Some Israelis argued in favor of taking the predictive test in order to be able to be better prepared in terms of life planning that is being conducted together with other family members. It was common among Israeli respondents, but not German respondents, to speak about their motivation for taking the predictive test in the context of their children—current and future. Israeli respondents also brought up prenatal testing in this context. In addition, among our Israeli participants, considerable support for telling family members about the test results was apparent across all groups. As explained in detail in Chap. 5, the majority of Israelis said that if they were to take the predictive test, they would share the results with family members. This argument stresses that in this situation there is no doubt that the children need to know as well—that if you already have children you should tell them so they will have the opportunity to be tested themselves. Indeed, arguments raised in favor of telling family members about test results included responsibility for one's spouse and kin in sharing relevant information for life planning. The major point was that the children should be made aware that they are at higher risk than the general population.

In Germany, only a minority—in all groups—supported the idea of sharing test results for either colon cancer or Huntington's disease with family members, as we already explained in Chap. 5. German respondents who spoke against telling their family members emphasized two different arguments: First, the anxiety that such information may cause for the family member, and second, the unwillingness to share what was strongly perceived as personal information. Few German respondents discussed in this context the "alternative" decision not to have children, or—in case one carried a gene for Huntington's disease—to adopt children. Hence, family responsibility in the case of hereditary disease can even lead to consider alternative models of a family. It does not necessary imply solely a biologistic ideal of family.

Once again, in relative terms, the largest support of telling family members was expressed in the affected German group. Some affected German respondents also spoke in a parallel manner as the Israeli respondents about the right to know of family members. This right to know of others corresponded with one's own duty to know and to tell, as part of their responsibility towards kin.

In the context of end-of-life care and advance directives, a major argument was commonly made across all groups, in Germany and Israel. This claim regards the association between responsibility of others and the risk of losing one's autonomy. In this context, responsibility was most obviously perceived as the responsibility of family and doctors towards preserving and protecting one's autonomy and dignity. The reason behind this conviction is that autonomy is considered an important and constitutive element of one's self. Participants in both countries agreed that a medical treatment should only be started if it is in accordance with the patient's will. The advance directive was seen as a way of expressing one's will even when that person can no longer speak up. The document should, hence, be respected in the same way as within an informed consent procedure or the socially deep embedded practice of one's last will. As we elaborated in Chap. 7, this argument stresses that for the sake of life with dignity, it should be the duty of doctors to implement people's requests. The idea of withdrawing an already started treatment like artificial ventilation caused more controversy driven by the risk of playing God, of compromising the sanctity of life to preserve the dignity of life.

The common view among Israelis regarding family members was that they are those who "naturally" should have the responsibility in terms of executing advance directives and acting in the best interest of the dying person. German respondents, in contrast, were more critical of the capacity of family members to speak for the patient. This argument stresses the difficulty due to the closeness of the relationship, differences in opinion amongst family members, and their inability to say goodbye. Thus, there are limits to family responsibility where family becomes a blurred concept. This is particular true in the German context, where pluralization and—as some would put it—the erosion of family increasing one's solitude with independence is seen as an ideal (Beck-Gernsheim 2000, 2006). This is a powerful cultural script that can offer a helpful analytical lens through which we can gain a better understanding of the cultural impacts on lay moralities. Yet another concern, which was discussed in the affected groups in both Germany and Israel, was that relatives might want to hasten the death of an elderly person because of claims to an inheritance. This argument indeed brings the risk associated with placing too much responsibility in the hands of relatives to an extreme.

To sum up, the strong claims as well as concerns considering a relative's role in respecting a family member's wish for end-of-life decisions (as well as respecting his or her wish to live as long as possible) can be interpreted in the broader context of "death" as social death: Dying and death are not only an individual experience but always strongly affect the close social environment. So asking a family member to hasten one's own death in the form of assisted suicide for many is considered a taboo because close family members will suffer death due to a beloved and grieving person. Also speaking of self-inflicted suicide is socially problematic and publicly uncommon—in contrast to the omnipresence of the topic in movies and literature. The focus group discussion functions as a "small public." It therefore mirrors what is expressible in a public discourse—and what cannot be said. A doctor assisting in the process of dying is for many more logical and acceptable, but conflicts with his or her professional responsibility to save lives. The legal restriction is interpreted as an expression of state responsibility to protect the doctor's integrity and responsibility to save lives. Hence, critical voices that radically demand a legalization of doctor-assisted suicide are only found in politically organized lay groups such as the constituency-based organizations *Dignitas* (Germany) and *Lilach* (Israel). Both are fighting for more patient rights to self-determination including the right to assisted suicide (Raz et al. 2014).

8.3 Responsibility of Society and the Risk of Social Coercion

Both advance directives and genetic testing are biomedical practices that have a potential connection to euthanasia and eugenics, concepts that carry loaded moral implications and historical associations. This might be particularly true for Germans and Israelis but also for many Western countries in which eugenics and euthanasia were embedded in local culture (see for instance Weindling 1993; McLaren 1990; Thomson 1998; Spektorowski 2004; Currell and Cogdell 2006). What stands out in our study is the direct reference made by so many German respondents to issues such as "social hygiene" in the context of genetic testing and "mercy killing" in the context of euthanasia. While "eugenics" functioned as a metaphor for general rejection of the idea of selection, dictatorship or racism based on genetics, there was also high sensitivity for other, more modern forms of genetic discrimination. For example, other German respondents also stressed the social and economic discrimination that might accompany too much genetic testing.

There was a consensus that genetic information should always be kept confidential unless otherwise decided by the patient. The state was considered responsible for this. However, the majority of Israeli respondents did not express mistrust of the state and of doctors, or worries regarding potential abuse, in the manner that characterized the Germans. On the contrary, Israelis had a positive view of state regulation and many agreed that the state and HMOs should finance genetic tests as part of public health care. Indeed, while many Israeli respondents argued in favor of a cost-benefit approach to state-managed genetic testing, they did not—in contrast to German respondents—consider any negative aspects of social coercion that might accompany such public health measures. This was rather exceptional since in many other aspects of social life such as national security, the economy, or education, Israelis are quite critical of the state. Similarly in the case of advance directives, Israeli respondents saw the role of the state not with suspicion of social coercion but with trust, for example, as purveyor of standardization.

Some among the affected German respondents expressed a view that drew on the German cultural grammar, however, in a way that was closer to the Israeli interpretation. They referred to the social memory of eugenics, but did it as something which is hindering progress. They argued that it is the responsibility of a welfare state to provide genetic testing on a regular basis. Finally, both German and Israeli respondents in the affected groups focused on their needs as affected people, justifying the involvement of the state that is required to protect their particular interests but also can be interpreted as a claim for a paternalistic state that needs to nudge ("start a campaign") or care for its citizens to protect them from laziness or ignorance. Affected German respondents expressed similar views to those of the Israeli respondents, stressing that the responsibility of society should be expressed in more regulation and not in allowing a market-driven or doctors' interest-driven laissez faire.

Being affected also resulted in more similarity across countries in the context of end of life. Whereas many German (non-affected) respondents were alarmed by the very mention of active euthanasia because of its historical abuse, arguments in support of active euthanasia were voiced by both German and Israeli affected respondents in a way that focused on individual autonomy and personal needs. The focus on autonomy was so strong that it appeared to overrule the general concerns regarding euthanasia and the risk of social coercion. This argument stresses, as we explained in Chap. 7, that active euthanasia is only acceptable in extreme cases—in medical conditions where there is no other chance of cure and the patient is fully conscious and responsive. The similar bottom line for those who are affected was that we should enable people to end their life when nothing else works. The crux of this argument was an assertion concerning the ultimate value of autonomy: "The life of a person is that person's life." The risk of losing that autonomy, hence, overruled the responsibility of society to regulate all other aspects of end-of-life care, including misuse.

8.4 Conclusions

We found three dominant themes of responsibility that characterized the discussion: self-responsibility, family responsibility, and state responsibility towards its citizens. These themes were furthermore matched with a corresponding perception of particular risks. What can these findings tell us about the complex ways in which different cultural grammars, as well as the common experience of being affected, influence the ways in which people make sense of risk and responsibility in health and illness?

In Germany and Israel, responsibility was associated with the potential individual risk of suffering from a genetic disease as well as the potential loss of individual autonomy. However, there were diverse cultural grammars which modified this basic argument. In Israel, supporting genetic testing if the risk could be prevented by treatment was a common argument across all groups and was expressed by the majority. In Germany, support of taking the predictive test even when a pre-emptive treatment existed was, in stark contrast, a minority view, although dominant for all those affected. This finding appears to reflect a significant cultural difference between lay people in the two countries. The Israeli respondents stressed the benefit of the test as leading to pre-emptive treatment; while the German respondents emphasized that treatment is available irrespective of the person's preknowledge about his or her genetic status, thus the test does not make a difference.

The respondents extended the discussion of responsibility to include family responsibility in the context of genetic testing as well as advance directives for end of life. Here, the diversity of cultural grammars also loomed large. In Israel, support for telling family members about genetic test results was expressed across all groups by the majority. In Germany, support of telling family members about the test results was a minority view in all of the groups; the majority emphasized the anxiety and psychological stress that such information may cause, and the unwillingness to share what was strongly perceived as very personal information. While health professionals encourage people to pass on genetic risk information to their relatives, the decision whether, when and how to do this is not trivial but a complex issue influenced by familial and cultural factors as well as individuals' responses to risk information (Keenan et al. 2005). The cultural difference found in this study could be connected to the family values which are characteristic of a family-oriented society such as Israel, versus the value of self-determination which is characteristic of a more individualistic society such as Germany.

This difference in cultural grammars could also explain the findings in the context of end of life regarding the responsibility of others' family members and the professional responsibility of doctors for respecting advance directives vis-à-vis the risk of losing individual autonomy and the risk of futile suffering. In Israel there was more support of the view that responsibility should lie first and foremost with family members. In contrast, German respondents argued that decision-making should be based on shared responsibility of doctors and family members. In Israel, current practice shows a relative lack of shared decision-making, also in the context of end of life (Miron-Shatz et al. 2011). For Germany, Härter et al. (2011) show that besides a variety of activities and training programs for shared decision-making work needs to be done for establishing shared decision processes as routine also in the context of end-of-life decisions. The issue of shared decision-making in end-of-life care is, hence, expected to continuously raise conflicts and deliberations.

Another difference in cultural grammars was the impact of religion. Jewish respondents argued against withdrawal from treatment and assisted suicide on religious grounds because of compromising the sanctity of life. This argument indeed also underpinned the Israeli law for the terminally ill patient (2005), where (in contrast to the German law), the application of advance directives is limited to terminally ill patients in the last six months of their life; only withholding (rather than withdrawing from) treatment is allowed; and advance directives are only recognized in a standard, long, jargon-loaded formal form issued by the Ministry of Health. This explains perhaps also why Israeli respondents trust their family more than the medical system.

Finally, responsibility of the state was connected to the risk of social coercion but only among German respondents, where the cultural grammar of the Holocaust and the Nazi regime loomed large. German respondents also stressed their mistrust of the economic interests of private insurance companies arguing that genetic information must remain extremely confidential otherwise it will become a basis for wide-ranging discrimination. In the context of euthanasia, too, many German respondents mentioned that the very term is alarming and related to potential abuse, expressing, for example, mistrust of doctors who might prefer to let somebody die so their organs can be transplanted. In contrast, the majority of Israeli respondents did not express mistrust of the state, or of doctors, or worries regarding potential genetic discrimination and abuse. On the contrary, there was a positive view in support of increased state regulation as purveyor of public health genetics and standard ADs.

We discuss the perspective of affected respondents separately in the ensuing section, after commenting on the broader cultural scripts underpinning these different cultural grammars. Against the contemporary Foucauldian perspective of bio-governmentality (Rose 2006; Beck 1992; Lemke 2011), our findings illustrate the need for further bottom-up differentiation. Responsibilization always is understood as a network of responsibilities, not as a single-actor issue. This more nuanced, culture-based understanding of responsibilities can be used to flesh out a concept that so far has remained very abstract in philosophy, and to develop this concept further on an empirical level. Because responsibility is always about our relationships with others, it is neither purely communitarian nor strictly liberal, but rather occupies a third, hybrid space of morality in-between these two opposites.

8.4.1 The Perspective of Being Affected

People affected by a chronic health condition can develop a common worldview of risk and responsibility that centers for example on the notion of endurance—of struggling with the disease rather than merely suffering from it (Honkasalo 2008). Being affected, in contrast to the abovementioned cultural grammars, was found to be connected with a line of arguments that provided more uniformity. This was particularly apparent in the context of genetic testing. The largest support concerning self-responsibility to take the test and to tell family members—which was the common view among affected as well as non-affected respondents in Israel—was also expressed in the German affected group. Affected German respondents also drew on the German cultural grammar regarding eugenics but in a way that was closer to the Israeli interpretation—arguing that the social memory of eugenics in Germany is hindering progress when it is conjured to raise public fear of genetic testing, while genetic testing should actually have been the responsibility of a welfare state. Finally, both German and Israeli respondents in the affected groups focused on their needs as affected people, justifying the involvement of the state so as to protect their interests.

Why was there such a contrast in Germany between affected and non-affected, but not in Israel? In Israel, as other studies have shown, "genetic anxiety" (or "responsibility," depending on one's perspective) has been constructed by professionals, experts and the public as a collective frame of risk in which the "elective" uptake of genetic testing is exceptionally high and seen by many as normative (Remennick 2006; Raz 2009). The moral argument regarding the duty to know reflects, in the case of Israel, a confluence of public and professional (medico-legal) worldviews promoting genetic testing as a collective agenda (Zlotogora et al. 2009). A collective lay notion of "being affected" could thus explain why the Israeli lay morality of responsibility is in fact a morality of being affected (Raz 2009).

In the context of end-of-life care, affected respondents in both Germany and Israel spoke about their concern that relatives might want to hasten the death of an elderly person because of financial reasons. There was an overall consensus, cutting across affected and non-affected respondents, regarding more general issues (such as advance directives and passive euthanasia). This overlapping consensus regarding the need to proceed with caution and strive for shared decision-making involving all stake-holders probably reflects the fact that the general public is increasingly affected by the greying of society and the media exposure of end-of-life dilemmas. Unlike genetic testing, where awareness may depend on one's family history, we are all affected by end-of-life questions, and living wills or advance directives are, generally speaking, relevant to everyone. The most significant argument reflecting the unifying perspective of being affected in the context of end-of-life care was the support of active euthanasia in extreme cases, where the risk of losing one's autonomy and dignity overruled the responsibility of society to regulate all other aspects of end-of-life care. In another study we similarly found that constituency-based patient organizations whose membership was mainly comprised of affected people, like *Dignitas* in Germany and *Lilach* in Israel, sought to change the existing medico-legal policies regarding euthanasia by advocating for "the right to live and die in dignity" (the acronym of *Lilach*), which they considered emancipatory. A large group of *Lilach*'s members are in favor of death by prescription (Raz et al. 2014). This stance contrasts the perspective of the general public and the legal policy regarding euthanasia in both Israel and Germany.

In sum, this study demonstrates that a multifaceted awareness of the variety of worldviews, including the attitudes of those affected by clinical symptoms or by being a carrier, and of those not affected, warrants attention by sociologists, health professionals, policy-makers, bioethicists, and all those who are interested in a more nuanced understanding of health, risk, and society. A focus on the interplay of cultural grammars as well as the perspective of being affected, and their combined effect on concepts of responsibility and risk, provides a helpful, although little explored interface for bridging some of the gaps between experts' formal ethics of principles and our lay moralities.

References

Ariès P (1981) The hour of our death. Knopf, New York

Arribas-Ayllon M, Sarangi S, Clarke A (2008) Managing self-responsibility through other-oriented blame: family accounts of genetic testing. Soc Sci Med 66(7):1521–1532

Beck U (1992) Risk society: towards a new modernity. Sage, London

Becker E (1973) The denial of death. Free Press, New York

Beck-Gernsheim E (2000) Was kommt nach der Familie? Einblicke in neue Lebensformen, 2nd edn. Beck, München

Beck-Gernsheim E (2006) Die Kinderfrage heute. Über Frauenleben, Kinderwunsch und Geburtenrückgang. Beck, München

Beier K, Jordan I, Schicktanz S et al (2015) Familien und Patientenorganisationen als kollektive Akteure in der Bioethik: vernachlässigt und unterschätzt? In: Steinfath H, Wiesemann C (eds) Autonomie und Vertrauen in der modernen Medizin. Springer, Heidelberg (in press)

Conrad P (1994) Wellness as virtue: morality and the pursuit of health. Cult Med Psychiatry 18:385–401

Currell S, Cogdell C (eds) (2006) Popular eugenics: national efficiency and American mass culture in the 1930s. Ohio University Press, Athens

Härter M, Müller H, Dirmaier J et al (2011) Patient participation and shared decision making in Germany—history, agents and current transfer to practice. Z Evid Fortbild Qual 105(4):263–270

Honkasalo ML (2008) Enduring as a mode of living with uncertainty. Health Risk Soc 10 (5):491–503

Keenan KF, Simpson SA, Wilson BJ et al (2005) "It's their blood not mine": Who's responsible for (not) telling relatives about genetic risk? Health Risk Soc 7(3):209–226

Lemke T (2011) Critique and experience in Foucault. Theory Cult Soc 28(4):26–48

McLaren A (1990) Our own master race: eugenics in Canada, 1885–1945. McClelland & Stewart, Toronto

Miron-Shatz T, Golan O, Brezis M et al (2011) The status of shared decision making and citizen participation in Israeli medicine. Z Evid Fortbild Qual 105(4):271–276

Nassehi A, Saake I (2005) Kontexturen des Todes. Eine Neubestimmung soziologischer Thanatologie. In: Knoblauch H, Zingerle A (eds) Thanatosoziologie. Tod, Hospiz und die Institutionalisierung des Sterbens. Duncker & Humblot, Berlin, pp 31–54

Raz A (2009) Community genetics and genetic alliances: eugenics, carrier testing, and networks of risk. Routledge, London

Raz A, Jordan I, Schicktanz S (2014) Exploring the positions of German and Israeli patient organizations in the bioethical context of end-of-life policies. Health Care Anal 22(2):143–159

Remennick L (2006) The quest for the perfect baby: why do Israeli women seek prenatal genetic testing? Sociol Health Illn 28(1):21–53

Rose N (2006) The politics of life itself: biomedicine, power, and subjectivity in the twenty-first century. Princeton University Press, Princeton

Rose N, Rabinow P (2006) Biopower today. BioSocieties 1(2):195–217

Spektorowski A (2004) The eugenic temptation in socialism: Sweden, Germany, and the Soviet Union. Comp Stud Soc Hist 46(1):84–106

Thomson M (1998) The problem of mental deficiency: eugenics, democracy, and social policy in Britain, c. 1870–1959. Oxford University Press, Oxford

Turoldo F (2009) Responsibility as an ethical framework for public health interventions. Am J Public Health 99(7):1197–1202

Weindling P (1993) Health, race and German politics between national unification and nazism, 1870–1945. Cambridge University Press, Cambridge

Wiesemann C (2010) The moral challenge of natality: towards a post-traditional concept of family and privacy in repro-genetics. New Genet Soc 29(1):61–71

Wöhlke S (2014) Lebendorganspende. Zur Relevanz von Geschlecht und Reziprozität. Wege zum Menschen 66(4):600–614

Wöhlke S (2015) Geschenkte Organe? Ethische und kulturelle Herausforderungen bei der familiären Lebendnierenspende. Campus, Frankfurt (in press)

Zimmermann C, Rodin G (2004) The denial of death thesis: sociological critique and implications for palliative care. Palliat Med 18(2):121–128

Zlotogora J, Carmi R, Lev B et al (2009) A targeted population carrier screening program for severe and frequent genetic diseases in Israel. Eur J Hum Genet 17(5):591–597

Chapter 9
Final Conclusion: Disentangling the Micro and the Macro in Bioethics

When lay people in Germany and Israel talk about responsibility concerning their own and other's health, do they just tell a personal story? Narration, we argue following many others (Gergen 1991, 1994; Shotter and Gergen 1989; Gubrium and Holstein 1998), is a dialectical process that combines micro and macro levels—social scripts and individual meaning. Such narration may take place in every interaction, and maybe even more so with heightened self-awareness in the social context of the interview or focus group. Ethnomethodology, for example, considers interviews as an ongoing, interpretive accomplishment, a collaborative construction of meaning (Garfinkel 1967; Heritage 1984) that furthermore imposes particular ways of passing and accountability (Cicourel 1964). This applies just as well, if not more, to group interviews such as focus groups. Seen from a symbolic interactionist perspective, interviews become "inter-views" (Raz 2005). Ethnographic analyses have similarly come to emphasize the interview as a site for dialogue where cultural members are seen as ethnographers in their own right (Atkinson 1990; Clifford and Marcus 1986). The task of disentangling the individual sense of morality from the social-collective principles of ethics that shape our stories is thus one of the major challenges this study is facing.

In establishing a given endpoint and endowing it with value, and in populating the narrative with certain actors and certain facts as opposed to others, the narrator enters into the world of moral and political evaluation. Value is placed on certain goals (e.g., "life with dignity"), certain human relationships (e.g., self-oriented as opposed to family-oriented), and particular modes of description (e.g., the world as material as opposed to spiritual). When our focus group participants narrated their claims, they could also be seen as reasserting some of the topoi which unite culture into a tradition.

We should be reminded once again about the social framing that the focus group constructs for its members. When a child tells her parents the story of her day in school, she is simultaneously constructing an image and sustaining a relationship; in the telling, her parents may be constituted as guardians, and the simple story may become an integral contribution to the sustenance of the family. A parallel process may take place in extended group discussions where participants reassert and sometimes challenge in the presence of others their shared systems of meaning, which is their culture.

A.E. Raz and S. Schicktanz, *Comparative Empirical Bioethics: Dilemmas of Genetic Testing and Euthanasia in Israel and Germany*,
SpringerBriefs in Ethics, DOI 10.1007/978-3-319-32733-4_9

Can we safely assume, then, that our participants were involved in telling together a "German" or an "Israeli" story? The "national story" was for a long time dominant in bioethics, because national legal frameworks and policies are closely connected with professional ethics (e.g., Hoshino 1997). In addition, the understanding of the "national story" was seen by some as a stepping stone for a "global bioethics" (e.g., Tao 2002). The interest in professional ethics is still an important point of departure for many scholars in bioethics. In this book, we subverted this dominant focus by looking not at professional ethics but at lay person's morality. As we recapitulate and look back at the findings, parts of the answer(s) begin to take shape, while others still remain to be examined. This is exactly the kind of inductive uncertainty which makes the sociological and the everyday life study of morality related to responsibility so difficult, but also so interesting.

It should be clear by now that we are not defending the idea that the socio-empirical analysis of morality will serve *eo ipso* as moral justification or normative orientation. Instead, our approach is located on the spectrum of descriptive yet self-reflective bioethics. Descriptive bioethics differs from normative bioethics in its utmost concern: while the former describes and classifies existing positions (the "how and who"), the latter gives reason, judges and provides advice for future actions (the "why"). As argued earlier, however, there are several approaches in normative bioethics which do not detach themselves from existing moral experience and moral practice, such as common-sense morality (Gert 2004), deliberative participatory political ethics, hermeneutic ethics (e.g., Van Tongeren 1996), or even virtue ethics (e.g., Macintyre 1984). While the differences amongst these approaches are still a matter of on-going scholarly debate, many of them neglect the obvious fact that we already live and act in a highly self-reflective culture. Our continuous practice of moral negotiation serves as a starting point where abstract norms, social practices, and personal experiences are balanced, adjusted and weighed against each other. Moreover, social science, ethics and political philosophy are already expressed in lay persons' perception and thinking.

Lindemann et al. (2009) have argued in a similar way for a so-called "naturalized bioethics" in which they propose that bioethics needs to be empirically nourished. The broad spectrum of approaches presented in their book illustrates the kind of contextualized, more applied, and more inclusive perspective that we also advocate. The term "naturalized," however, is somewhat misleading as it covers up the social construction of such bioethics and the norms it addresses. The adjective "naturalized" may be read as a concession for physicians and biomedical researchers that may be tired of and overwhelmed with philosophical abstract theories. The desire to reconstruct an ethics that is closer to human "nature" may however be too reductionist. Furthermore, the contextual nature of many of these approaches lacks systematic comparison that can make us aware of our local blind spots and implicit premises. Emphasizing the local and contextual nature of the problems of bioethics is not enough; it needs to be explored and compared so that common and shared issues can also come to the surface.

We agree absolutely with Lindemann et al. (2009) overall endeavor, but we would like to add to it comparative cross-cultural bioethics as a crucial element that helps close down the gap between expert and public culture. Our aim is not to deduct from such cross-cultural observations whether the "right not to know" in genetics is justifiable or if assisted suicide should be allowed or forbidden, because it seems a consensus in the lay sphere. Rather, our aim is to add these perspectives into the discussion, challenging the dominance of professionals, experts and politicians.

As we stressed earlier, our purpose was not to discredit established avenues. Our aim is rather to propose a new perspective, that cross-cultural comparative bioethics can be matched with all the above mentioned approaches. It is born out of a dialectical understanding of different strands in moral thinking. It respects, on the one hand, the shared human experience of specific, context-related and experience-loaded moral statements. On the other hand, it acknowledges also the reasonable underpinning that social normative orientation goes beyond an individual opinion and strives for joint commitment, a collective agreement and social acceptance of norms beyond the very local context. This approach is particularly useful for heterogeneous, postmodern societies and groups where moral pluralism (rather than moral consensus) is more likely to prevail.

Our approach is built upon a reflective model of empirical ethics (Holm and Irving 2004) in which empirical knowledge of the practice informs directly the reflection on the ethical relevance of hypothesized problems. It is rooted in the philosophical methodology of reflective equilibrium for applied ethics (van Thiel and van Delden 2010; Daniels 1979, 1996; Ebbesen and Pedersen 2007). This methodology assumes that in a pluralistic social setting, abstract ethical considerations must be revised on the basis of commonly shared values or moral judgments. Empirical insights substantiate much better than a philosopher's armchair thinking what is a "considered moral judgement" (Elgin 1999; Hahn 2000). Situated knowledge and experience of the persons affected serve for refining general ethical

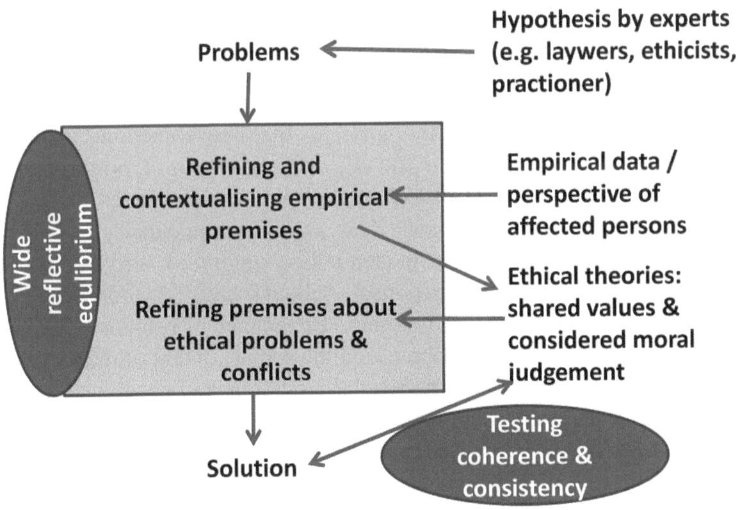

Fig. 9.1 Scheme of a systematic, deliberative approach based on a wide reflective equilibrium including theoretical as well as context-sensitive perspectives

rules and social expectations. It allows, at least partly, for a reflective juxtaposition of the macro and the micro, which through democratic deliberation may lead to the formulation of socially accepted directions for future practices (see Fig. 9.1).

The consideration of persons being involved and affected can be justified by the so-called participatory approach (see Schicktanz 2009b, 2012; Schicktanz and Schweda 2015). It starts from participatory considerations based on Jürgen Habermas' deliberative ethics but also takes into account socio-practical objections. Such critique, as formulated by Iris Young or Nancy Fraser (Young 2002; Fraser 1989), can be summarized in the following manner: The original discourse of ethics may be too naïve concerning the impact of social power, and neglects vulnerabilities and the omnipresence of power and injustice (Fricker 2009). To address these concerns, affected persons (incl. patients, family members and care takers) must be included not only in the process of deliberation and decision-making, but already accounted for in the basic formulation of which problems need to be solved.

As argued earlier, responsibility is a meta-ethical principle allowing for systematic consideration of relational, contextual and empirical insights. It overcomes accepted deductive distinctions such as self/other, subject/object, and doctor/patient because it asks always for the underlying (theoretically and/or empirically underpinned) relationship. Sensitivity for relational conditions means also to critically question these labels. When are doctors "experts" and not also/just lay persons? When does a patient become an "expert"?

Empirical ethics, for example in the form of narratives about moral experience, takes as its starting point the individual life experience in its unrepeatable uniqueness. Therefore it resists tendencies to generalize, even at the risk of being accused of theoretical inadequacy. Can there be a bridge between the two? Any analytical attempt to distill a "national" experience or some other thematic principle runs the risk of over-generalization as well as contextual dilution. Before trying to apply any general formulae to the experiences, attitudes and narratives we collected, several caveats must be spelled out and considered. First, most of our data consist of attitudes toward given scenarios, rather than of complete stories or narratives. Some of these attitudes are expressed within narratives that may sometimes take the form of stories, with protagonists (subjects and objects), motivations ("norm supervising" within a subjects' moral context of legitimation), plot (consequences, sanctions) and chronotopes (space and time frames). This is not a coincidence: when we are speaking about responsibility we are indeed telling a story—a story about someone who "should be able to provide a response"—about what? To whom? This already hints at the situated, contextual and empirical character of responsibility (Keenan 1997). Such narration takes a special place in the social context of the focus groups. When our focus group participants narrated their claims and positioned them in the group's discourse, they could also be seen as reasserting and sometimes challenging in the presence of others (a "mini-public") their shared systems of meaning, which is their culture.

In addition, the attitudes and stories we focus on as evidence for ideational patterns of a popular or common culture, it should be noted, are treated primarily outside of their immediate sociocultural contexts. Particular communities and speech situations are hence underemphasized. The focus group situation is artificial, as well as the use of vignettes as triggers for the conversation. Narratives are thus treated here more as informational texts than as interactional, and indexical, performances; the particular histories of tellers and tellings, the subtleties of situated tellings, are to a large extent elided.

In what follows we focus on a few stories that were told by participants in order to substantiate their claims. We focus on these narratives/stories because they also substantiate general patterns that were found to characterize the groups in terms of their national, religious and affected dispositions. Each of these stories is unique and personal, yet it also has a deep structure, conveying some of the group's values and norms. Our ability to draw the connections between the story and the script, the personal and the communal, builds on the accumulative knowledge we have gained from working on these narratives, analyzing and comparing them. Only following this long inductive process we can connect the personal contours to their underpinning, deep structure schemes. It is hence appropriate to do this in the conclusion as a way of juxtaposing the micro and macro levels of our overall analysis. In doing such a detailed analysis of a few stories, we also realize that there is no such thing as "one story"; each story contains, corresponds with, and is made up of multiple references to other stories. No story exists in itself; no story is an island. There is no one single "affected story" as there is no one single "German" or "Israeli" story. Each story is like an endless maze where each turn, each writerly reference as well as readerly interpretation, can immediately take you to a new hermeneutic maze.

> By the same process whereby Man spins language out of his own being, he ensnares himself in it; and each language draws a magic circle around the people to which it belongs, a circle from which there is no escape save by stepping out of it into another (Wilhelm von Humboldt, quoted in Cassirer 1946, p. 9).

The following narrative was told by a German male in the affected group. It contains self-reflection on complementary German worldviews, as well as criticism of them, from the perspective of being affected by a genetic disease. In that sense, affected people can be regarded as the "other within us," as anthropological observers of their own society. The narrative starts with a general outlook and then becomes personal:

> I believe that in the case of hereditary diseases, this is progress. If people are made for each other, they shall have, or shall not have children. I think that this is a responsibility for the common welfare that, in my opinion, is nowadays not seen very often in fear of reverting back to Nazi eugenics. But there is responsibility. In this respect, I deem genetic diagnostics at this point as responsible.

This short narrative exposition combines, in a nutshell, some deep-seated German outlooks and the self-reflection of being affected. We have found that by and large, perceived from the perspective of being affected, genetic diagnostics of hereditary diseases is good; "this is progress." This is a utopian view implicitly rejecting the dystopian outlook of some disability groups in Germany (for the German debate see, e.g., Köbsell 2003; Graumann 2003; Lenhard 2003; in general also: Johnstone 2001), being concerned with the use of genetic testing as potentially leading to a slippery slope of discrimination, and the selection of life along the yardstick of which form of life is worth living, being "very often in fear of reverting back to Nazi eugenics." The perspective of being affected hence appears to be expressed in more pragmatic terms, being critical of overly prudential ideas, although it remains critically grounded and alert—"I deem genetic diagnostics at this point as responsible."

By addressing the argument whether there is a risk of "liberal eugenics" (Habermas 2003) or not, the speaker positions himself in an ongoing, "typical" German discourse.[1] He is not simply referring to his own suffering, quality of life or frustration over a doctor. Instead, the speaker stresses a general claim that is very consensual in current German culture on having children: "If people are made for each other, they shall have, or shall not have children"—the legitimacy of being childless out of your own individual choice being characteristic of the German outlook but not, as we saw, of the Israeli one. Having spelled out some of the major cultural underpinnings of this narrative, let us continue with its more personal part. Evidently the two parts are inter-connected.

> But the personal part is: A year ago I was given the surprise diagnosis of cancer in an advanced stage. My reaction was and still is important to me: I was not shocked and thought: Well, this is a reality with which I have to live. I never expected it. And then, there it was; and the moment I heard it, it became a fact to live with. This reaction of mine is important to me. It is very valuable. The spontaneity that I had kept by not knowing that there might be a danger. And then this possibility to react within this spontaneity, in the way that is deemed appropriate to me. This somehow strengthened my character.

This part of the narrative illustrates the significance of the "illness narrative" for the on-going formation of self-identity. As the illness (the subjective phenomenology of the disease) becomes one with the person, it changes one's sense of personal identity. Evidently, this process works differently for different people. For some, the realization can be a source of strength, for others—a source of weakness. Some may advocate the duty to know what is in our genes, while others may uphold the right not to know about a future disease. This is exactly why there is no single "affected story," but rather a plurality of stories, which may furthermore change over time with individual changes. Some would argue that such hermeneutics amounts to an epistemological impasse in translating one materiality to another due to unshareable properties and through lack of constructible commonalities. Our Western, midlife norms and expectations prescribe that identity will be communicative and dialogical at moments of expression, listening and speaking. The incompatibility of space and time between healthy and ill, able-bodied and disabled, may upset the plausibility of managing otherness (Fabian 1983).

[1]We say here "typical German discourse," because as comparative research has indicated, other lay groups (outside of Germany) rarely mention the historical reference to Nazi medicine or Nazi eugenics (Felt et al. 2010). This implies, however, not that eugenic practice is not addressed in other countries, but the reference to historical practice is particularly significant for German debates.

However, as this book has shown, it is possible to understand the voices of those who are affected if we first lend our ears to them.

The following narrative was told by an Israeli woman (not affected). In Israel, we have found out, there is a common positive and even utopian outlook of biomedicine in general and of genetic testing in particular. This outlook, which we found primarily amongst affected people in Germany, is shared by the Israeli majority, by both people who are affected and people who are not. In a very similar manner to the previous narrative of the affected German respondent, the following story begins with a more general exposition that leads to a personal account:

> I guess in my youth I had more conservative opinions, like my parents. I believed in fate… and was more traditional. Today, with more advanced technology, and medicine and science having achievements that are so impressive, including for life and death, I believe that you can prevent much suffering by medical technology in general, and genetic tests in particular. I can say so myself: in my twenties, when I married, I was much less aware of the issue of genetics and its importance, unlike how I see it now. Therefore, I believe that I would not be going to do genetic testing even if it were possible then. I suppose that this change is a function of time and medical advances and technology. Today, I believe in medicine and in its ability more than ever, this is because I saw with my own two eyes how medical tests save lives.

On a personal level, this is a concise autobiographical narrative conveying an experience of striking insight for the female speaker. She contrasts her youth, when she was not aware of or even objected to genetic testing, to the present, in which (already in her 50s) she has become an avid supporter of genetic testing. This story of revelation is unfolded through a personal transformation from "traditional conservatism" to a belief in "medical advances," as the speaker puts it. This personal narrative represents the well-known, perhaps even dominant, modern, western story of medical progress. In 1998, Nobel laureate James Watson, who helped established the Human Genome Project, stated in a lecture, "We used to think out fate was in the stars. Now we know it is in our genes." This statement is also echoed in the speaker's words.

In this personal narrative, the value of genetic testing is not just abstract (medical progress) but also concrete and utilitarian: "I believe that you can prevent much suffering by medical technology in general and genetic tests in particular." In the context of the focus group discussion, this speaker went on to tell the group about people she knew when she was young, who were born with terrible genetic diseases that made them suffer a lot, and how such suffering has been prevented more recently with the advent of genetic testing. We hence find here the moral argument, made by many lay people in Israel, concerning genetic responsibility. Using genetic testing to prevent suffering is increasingly propagated and accepted as part of one's parental responsibility. This story could also be seen to convey a national script promoting genetic testing in the context of a collective sense of risk (Raz and Schicktanz 2009a, b). We do not find in this national script criticism of the technocratization and instrumentalization of medicine, as in the horrendous case of the Nazi Doctors used in slippery slope arguments by German experts and lay people. Rather, the "Israeli" story of lay people and experts alike links the utopian view of science and medicine to their role in the service of national survival (Prainsack and Firestine 2006; Prainsack 2006; Raz and Schicktanz 2009a, b).

In the context of end-of-life care, too, we found narratives that highlight the change of perception that is brought about by being affected. For both German and Israeli respondents, the perspective of being affected by the death or terminal illness of a close relative was characterized by a renewed sense of the need to respect one's autonomy even if it means letting go and withdrawing/withholding medical treatment. Many affected respondents commented how when they were young, their views were different—they believed that preserving the life of every human being was the most important thing. However, affected respondents in Germany and Israel commented how due to their own experience with their parents, their views began to change radically. As one of our affected Israeli respondents (female) put it typically:

> You could see the continuous suffering and the pain and you could hear her anguish. The will of the patient should be followed, if the patient prefers not to receive medical treatment, out of consideration for the autonomy of the patient.

A new form of responsibility was mentioned by some of the respondents in this context—the responsibility of parents to have advance directives so that the burden of decision making will not fall on their children alone.

Two important findings can now be highlighted concerning how lay people perceive end-of-life care. First, we found an overall consensus amongst all groups of lay people regarding the need to respect the autonomy of the terminally ill. Second, an overarching consensus was also found in terms of the perception of end-of-life decision making as an on-going, dynamic, and mutual responsibility. While affected people were perhaps more expressive and detailed concerning their change of attitude regarding the need to respect the autonomy of the terminally ill, we found this perception amongst many non-affected persons as well. There was an overall consensus, cutting across affected and non-affected respondents, in supporting passive euthanasia when it is in line with the ill person's will. Moreover, we found also overlapping consensus regarding the need to proceed with caution and strive for shared decision-making involving all stake-holders in end-of-life situations, reflecting the fact that the general public is increasingly affected by the greying of society and the media exposure of end-of-life dilemmas. Advance directives are, generally speaking, relevant to everyone and thus lay positions stress the importance of autonomy in the same way as experts do (Gedge et al. 2007). The most significant difference reflecting the perspective of being affected was the support of active euthanasia in extreme cases, where the risk of losing one's autonomy and dignity overruled the perceived responsibility of society to regulate all other aspects of end-of-life care.

The main difference between experts, doctors as well as ethicists, and our lay participants, is how end-of-life decisions were conceptualized. Focus group discussions started off with a binary question, e.g., withholding/not withholding treatment, which was later in the discussion framed by the respondents as a longer process of a series of decisions through which they accompanied (or would have liked to accompany) their relatives. As the abovementioned respondent typically put it: "We consulted in the family, between brothers, about this matter. All consultations were made with a doctor; we weighed the risks and benefits." This contrasts the often a-contextual framework of end-of-life decisions as unitary dilemma—construed as one moment of decision-making, which is used in ethical deliberations as well as in empirical research such as the EURELD study (Van der Heide et al. 2003). The prominence of so-called "advance care planning" emphasizing it as a continuous process shows parallels to this procedural understanding and should be also understood as a procedure taking place in different locations over a period of time: On the one hand, as continuous communication with doctors and care professionals (Cantor 2004), on the other within families (Schicktanz 2009a). It is important to realize that lay persons' descriptions of shared decision making in end-of-life care goes beyond the classical model of doctor-patient communication (Emanuel and Emanuel 1992) by conceptualizing the situation as one of shared responsibility. In end-of-life care, lay perceptions of the responsibilities of doctors are shot through with paradox: On the one hand, the doctors' moral source of authority is legitimized as part of their responsibility to save life. On the other hand, however, in end-of-life care this very responsibility is seen to limit and decrease the doctors' source of authority. This is the basis for the lay call for mutual responsibility in decision making, as lay perspectives are seen to have equal standing to medical expertise when it comes to determining the meaning of life with dignity and dying with dignity.

Despite the commonly shared lay perception of procedural and shared decision-making, this ideal is challenged by different cultural underpinnings of what a "family" means and how important it is. Many German lay persons pointed to the possibility that patients might have no relatives or none with whom they are on good terms. In contemporary Germany aging without children is indeed one of the most prominent topics (Kreager and Schröder-Butterfill 2005). In Israel, there were no such doubts that family members would not be there for the patient. This mirrors a more close-knit norm regarding the importance of the family for social life. This family ideal inter-connects the underlying pronatalist norm of creating a family (Birenbaum-Carmeli 2010; Hashiloni-Dolev and Shkedi 2007) with an underlying norm of dying within a family. Whether this fits always into reality with a growing number of singles living in modern Israel must be re-examined in the future.

What can be said about the influence of religion? While such influence is part of the multifarious parameters of the personal situation of the speaker, we could distill a few themes that reflect the different influences of Judaism and Christianity. The following narrative, by a German, not affected, modern-religious male, encapsulates the sense of prudence, also found among other German religious respondents, which emphasizes "not getting carried away" by new medical technologies, both in the context of predictive genetic testing and end-of-life care. Regarding these new technologies as verging on dystopia, the typically "Christian" German story advises to exercise caution:

> I believe that from an ethical point of view the topic whether or not [predictive] genetic testing should be allowed is very debatable. I would say no. I think prevention and the like; we have all that. Here in Germany, in Europe, we have reached a perfect medical standard. We can treat cancer, breast cancer. We can manage prostate cancer. We can control many types of cancer. We don't have to get carried away. One does not know scientifically what becomes of genetically manipulated maize. How it impacts the body. The food. I think, we as human beings, have to eventually admit to limits.

Interestingly, the speaker goes on to connect, within the same "Christian" narrative of prudence and self-restriction, predictive testing and end-of-life care:

> Yesterday, I was allowed to experience a funeral; that of my father. I am a Christian by the way. Well, not by the way, but… When I then… something became clear to me - I, too, worked in a hospice - that life ends at one point and that too is good. We should leave it that way, I think. We have a huge problem in the world. That is, people get older and older because of medicine. Some already ask how we will finance this. And one does not know what consequences that will have. One has to pay careful attention to that. And where are the limits. Where do we go after the genetic test? Yes. What comes next? No idea. Well, I would like to answer in the negative.

It is telling that the speaker himself connects his perceptions of prudence to his Christianity. In his words: "I am a Christian by the way. Well, not by the way, but…" Indeed, the same principle of prudence led some Christian ethicists to condemn genetics as interfering with processes (e.g., procreation and gestation) that have traditionally been under divine providence alone (Deane-Drummond 2006). Evangelical theologian Stanley Hauerwas highlights the importance of not interfering with the natural course of dying by claiming that "the most decisive challenge which medicine raises for Christian convictions and morality involves the attempt to make suffering pointless and thus subject to elimination" (Hauerwas 1986, p. 24). Interestingly, reformistic, modern religiosity often relies less and less on strict traditional everyday norms (e.g., what to eat, how to dress, what to do in particular situations; e.g., Bartmann 2003). Modern religiosity requires instead more self-reflective, self-control and self-responsible attitudes to decide what exactly it means to be a "believer." Hence, these overall attitudes seem to end in a more general attitude toward modern technology that can be labeled as "setting limits." It does not radically reject modern technology but problematizes the unbound and unleashed power of modernity embodied in modern technology, criticizing the automatism of following each innovation. On this level, we observed some similar, general, patterns of "hesitance" between modern religious Christians and Jews.

Although it would be tempting to look for a complementary "Jewish" story that emphasizes the benefits of biomedicine, this is only partially the case. The "Jewish" story, as told by our modern-religious respondents, advocates for predictive testing that carries benefit for the health of the individual in terms of preventive measures. When predictive testing is done by an adult for the sake of preventing or taking care of genetic diseases, this is regarded by Judaism as making a positive effort and fulfilling the mitzvah (religious prescription) of *hishtadlut* (Rosner 1991). Making an effort is connected with self-responsibility; the Jewish individual is expected to act responsibly not only towards themselves but also towards their children, including those not yet born. The Jewish bioethicist Brody (2002) demonstrates this stance when arguing that a couple has at least a minimal duty to use genetic testing to reduce suffering. For example, if a couple knows they are carriers of a severe disease, they should pursue genetic testing to prevent the birth of a child affected by that disease. This is also in line with the theological/conventional view of Judaism as supportive of medical efforts to intervene in nature for therapeutic reasons, seeing it as an appropriate use of the powers with which God endowed mankind (Wahrman 2002; Gross and Ravitsky 2003). The Christian "stewardship" model, in contrast, might be opposed to extreme medical intervening as "playing God" (Deane-Drummond 2006; Walter 1999). However, in the context of the end of life, the "Jewish" story was much less supportive of active interventions, prioritizing the sanctity of life over autonomy. This is exemplified in the following religious narrative, taken from an Israeli Jewish man:

> Disconnecting – that's something else. You all probably know this story, of a woman that was dying of some lingering disease, and her husband and son were trying by every means – including prayers in the synagogue – to keep her alive. She called them to her bedside and said that she was grateful for their efforts, but asked that they please refrain from such prayers because her life was no longer bearable. The rabbi was asked whether this would be permitted, and he answered that to refrain from praying is permitted, but that nothing positive could be done to shorten her life. The law in *Shulchan Aruch* states that you may not even move a pillow from underneath the head of a dying person in order to hasten his death.

For modern-religious respondents in both countries the difference between withholding and (actively) withdrawing treatment was much more important than for the lay respondents in the secular and affected groups. This supports the results of the comparative quantitative ETHICUS study (Sprung et al. 2007) regarding the difference between religious and secular doctors' perceptions of withdrawing life support. However, it is important to note that the ETHICUS study found a diversity of attitudes amongst protestant/catholic doctors depending on whether they were working in South, central or North Europe. This highlights once more the claim that there is no single "religious" story, but that cultural adaption is a strong force in the globalization of medical practice related to norms and morality.

To sum up, both the Israeli and the German narratives speak about the relationship between advances in health technology and increasing responsibilities. Before there were any genetic tests, one could be "traditional" in the sense of not knowing one's future and not being able to plan in advance. But to keep up with modern biomedicine, one should be aware of the technology and use it to one's benefit. Ignorance or passivity might prove to be, and condemned as, irresponsible. Testing in advance for late-onset diseases, as well as making plans in the form of advance directives for end-of-life care, are seen as increasing personal choice.

Comparative research is so important because it can serve as a translation process between everyday life plans and scholarly analysis of that "choice." We do not end up with concrete moral recommendations as to what to do—except for the recommendation to illuminate the individual dimension (for respecting autonomy) as well as the underlying group concerns and socio-cultural scripts.

What is the book's overall contribution to other areas in bioethics, especially those advocating a normative point of view? A major advantage of comparative cross-cultural bioethics can be seen in its pragmatic argument: It preserves a difference in perspectives which is otherwise easily overlooked by hegemonic discourses dominated by political or experts' voices. Our comparative approach aims at exploring cultural differences and moral plurality. However, it also allows for detecting overlapping areas. The latter is important for the production of socially robust knowledge and robust social decisions dealing with challenges induced by new biomedical technologies (Nowotny et al. 2001). For such an endeavor, considering interest groups such as patient organizations or other representatives and advocates of lay and affected people will contribute to a democratic socio-political decision-making process (Mansbridge et al. 2010). This is a serious aim given the fact that in various fields of life science research (e.g., genetically modified food or animal testing in neuroscience), a new kind of public "warfare" is going on, replete with conflicts and casualties.

Even as partisans of patients' and lay perspectives, we are not ignorant of the reproach of partiality: There is always a risk, in advocating the perspective of "being affected," to take too easily the side of those interviewed. If empirical bioethics focuses on too homogenous groups there is a real risk of producing too narrow views: the interviewed patients might be only interested in their own disease and situation. In so doing, they might neglect broader social issues that cause dilemmas for others. These worries are often expressed regarding patients participation in health services allocation, for example. Our findings do not support this worry, although they do not refute it; we have observed many cases where lay persons are also aware of third parties' interest. Comparative research confronts such risk through its built in juxtaposition of different groups thus ensuring a plurality of viewpoints. Different people, after all, can be differently affected. Furthermore, our proposed approach to enhance "inter-views" and to construe dialogues can at least methodologically circumvent the bias of personal stories. As the comparison between lay and professionals also illustrated, professionals and experts are not morally or epistemically neutral or objective. They are also deeply embedded in the macrostructures of the socio-political sphere. Hence, we defend our approach as opening up instead of closing down (Stirling 2008) our bioethical deliberations.

To avoid one-sided narratives, we seek to include the moral, affective and social dimensions of power by collecting and reflecting especially on stories of vulnerability, resistance, and alienation. Furthermore, innovative research in comparative bioethics should not only focus on mainstream publics, but especially consider and collect counter-narratives of counter publics. In this way subversive voices such as those of various advocacy groups (disability advocacy is a recent example) can be heard alongside mainstream medicine or bioethics. Newly emerging areas of such subversive advocacy include pre-demented persons that may have a conflict with relatives concerning life plans. Another case might be "savior siblings" (conceived through pre-implantation genetic diagnosis) or persons on the high functioning end of the autism spectrum.

How can we avoid "simple" stories, a difficulty many case-by-case morality approaches face? We suggest not to use stories as argument to defend a particular moral position, but to think within and through these stories: What is the hidden morality, how is that position exclusive, how ambivalent is the story teller herself? Sharing stories in public conversations and deliberations should be seen as translation and transgression of the "inner" logic, an opening into a complex process of interpretation which can enfold mutual understandings.

Another common argument in favor of professional ethics is that lay persons are at the end of the day not really in charge. Only doctors and politicians need to take responsibility for decisions. But this is not true. Patients and lay persons have to bear the consequences of policy decisions all the time. Furthermore, if policy-makers claim that individuals need to act self-responsibly, it also means these individuals should have the right to self-determination, also in the political realm. The flipside of proactive "self-responsibility" is the procedural right of taking part in decision-making. Responsibilization is not a one-way road as often claimed. We would like to stress that many dilemmas in bioethics are beyond the doctor-patient dyad. Rather they have broader socio-political and ethical implications of justice that cannot be left to the secluded circle of ethics committees or clinical ethics.

References

Atkinson P (1990) The ethnographic imagination. Routledge, London

Bartmann P (2003) Physician-assisted suicide and euthanasia: German Protestantism, conscience and the limits of purely ethical reflection. Christ Bioeth 9:203–225

Birenbaum-Carmeli D (2010) Genetic relatedness and family formation in Israel: lay perceptions in the light of state policy. New Genet Soc 29(1):73–85

Brody B (2002) Freedom and responsibility in genetic testing. Soc Philos Policy 19(2):343–350

Cantor MD (2004) Advance care planning in long-term care facilities. J Am Med Dir Assoc 5(22): 73–80

Cassirer E (1946) Language and myth. Harper & Brothers, New York

Cicourel A (1964) Method and measurement in sociology. Free Press, Glencoe

Clifford J, Marcus G (eds) (1986) Writing culture. University of California Press, Berkeley

Daniels N (1979) Wide reflective equilibrium and theory acceptance in ethics. J Philos 76:21–46

Daniels N (1996) Wide reflective equilibrium in practice. In: Sumner LW, Boyle J (eds) Philosophical perspectives on bioethics. University of Toronto Press, Toronto, pp 96–114

Deane-Drummond C (2006) Genetics and Christian ethics. Cambridge University Press, Cambridge

Ebbesen M, Pedersen BD (2007) Using empirical research to formulate normative ethical principles in biomedicine. Med Health Care Philos 10(1):33–48

Elgin CZ (1999) Considered judgment. Princeton University Press, Princeton

Emanuel EJ, Emanuel LL (1992) Four models of the physician-patient relationship. JAMA 267(16):2221–2226

Fabian J (1983) Time and the other: how anthropology makes its objects. Columbia University Press, New York
Felt U, Fochler M, Winkler P (2010) Coming to terms with biomedical technologies in different technopolitical cultures: a comparative analysis of focus groups on organ transplantation and genetic testing in Austria, France and the Netherlands. Sci Technol Human Values 34(4): 525–553
Fraser N (1989) Unruly practices: power, discourse, and gender in contemporary social theory. University of Minnesota Press, Minneapolis
Fricker M (2009) Epistemic injustice: power and the ethics of knowledge. Oxford University Press, Oxford
Garfinkel H (1967) Studies in ethnomethodology. Prentice Hall, Englewood Cliffs
Gedge E, Giacomini M, Cook D (2007) Withholding and withdrawing life support in critical care settings: ethical issues concerning consent. J Med Ethics 33:215–218
Gergen KJ (1991) The saturated self. Basic Books, New York
Gergen KJ (1994) Realities and relationships. Harvard University Press, Cambridge
Gert B (2004) Common morality: deciding what to do. Oxford University Press, Oxford
Graumann S (2003) Sind "Biomedizin" und "Bioethik" behindertenfeindlich? In: Akademie für Ethik in der Medizin e.V. (ed) Behinderung und medizinischer Fortschritt. Dokumentation der gleichnamigen Tagung vom 14–16 April 2003 in Bad Boll. Göttingen, pp 71–81. http://wwwuser.gwdg.de/~asimon/bb_2003.pdf. Accessed 30 Jul 2015
Gross ML, Ravitsky V (2003) Israel: bioethics in a Jewish-democratic state. Camb Q Healthc Ethics 12:3247–3255
Gubrium J, Holstein J (1998) Narrative practice and the coherence of personal stories. Sociol Quart 39(1):163–185
Habermas J (2003) The future of human nature. Polity Press, Cambridge
Hahn S (2000) Überlegungsgleichgewicht(e). Prüfung einer Rechtfertigungsmetapher. Alber, Freiburg
Hashiloni-Dolev Y, Shkedi S (2007) New reproductive technologies and family ethics: pre-implantation genetic diagnosis for sibling donor in Israel and Germany. Soc Sci Med 65(10): 2081–2092
Hauerwas S (1986) Suffering presence. University of Notre Dame Press, Notre Dame
Heritage J (1984) Garfinkel and ethnomethodology. Prentice Hall, Englewood Cliffs
Holm S, Irving L (2004) Research ethics committees in the social sciences. In: Kempf-Leonard K (ed) Encyclopedia of social measurement. Academic Press, London, pp 397–402
Hoshino K (ed) (1997) Japanese and Western bioethics. Kluwer, Dordrecht
Johnstone D (2001) An introduction to disability studies, 2nd edn. Fulton, London
Keenan T (1997) Fables of responsibility: aberrations and predicaments in ethics and politics. Stanford University Press, Stanford
Köbsell S (2003) Die aktuelle Biomedizin aus Sicht der Disability Studies. In: Schicktanz S, Tannert C, Wiedemann P (eds) Kulturelle Aspekte der Biomedizin. Campus, Frankfurt, pp 160–186
Kreager P, Schröder-Butterfill E (eds) (2005) Aging without children: European and Asian perspectives on elderly access to support networks. Berghahn Press, Oxford
Lenhard W (2003) Der Einfluss pränataler Diagnostik und selektiven Fetozids auf die Inzidenz von Menschen mit angeborener Behinderung. Heilpädagogische Forschung 29(4):165–176
Lindemann H, Verkerk M, Walker MU (2009) Naturalized bioethics: toward responsible knowing and practice. Cambridge University Press, Cambridge
Macintyre A (1984) After virtue: a study in moral theory. Duckworth, London
Mansbridge J, Bohman J, Chambers S et al (2010) The place of self-interest and the role of power in deliberative democracy. J Polit Philos 18(1):64–100
Nowotny H, Scott P, Gibbons M (2001) Re-thinking science: knowledge and the public in an age of uncertainty. Polity Press, Cambridge
Prainsack B (2006) Negotiating life: the regulation of embryonic stem cell research and human cloning in Israel. Soc Stud Sci 36(2):173–205

Prainsack B, Firestine O (2006) "Science for survival": biotechnology regulation in Israel. Sci Publ Policy 33(1):33–46
Raz A (2005) A note on inter-viewing: using symbolic interactionism for interview analysis. Stud Symbolic Inter 28:323–339
Raz A, Schicktanz S (2009a) Lay perceptions of genetic testing in Germany and Israel: the interplay of national culture and individual experience. New Genet Soc 28(4):401–414
Raz A, Schicktanz S (2009b) Diversity and uniformity in genetic responsibility: moral attitudes of patients, relatives and lay people in Germany and Israel. Med Health Care Philos 12(4): 433–442
Rosner F (1991) Modern medicine and Jewish ethics, 2nd edn. KTAV, Hoboken
Schicktanz S (2009a) Zum Stellenwert von Betroffenheit, Öffentlichkeit und Deliberation im empirical turn der Medizinethik. Ethik in der Medizin 21(3):223–234
Schicktanz S (2009b) Interpreting advance directives: ethical considerations of the interplay between personal and cultural identity. Health Care Anal 17(2):158–171
Schicktanz S (2012) Epistemische Gerechtigkeit. Sozialempirie und Perspektivenpluralismus in der Angewandten Ethik. Dtsch Z Philos 60(2):1–15
Schicktanz S, Schweda M (2015) Inklusive Deliberation: Zur Einbeziehung von Bürger- und Betroffenenperspektiven in medizinethische und gesundheitspolitische Entscheidungsprozesse. In: Rauprich O, Jox R, Marckmann G (eds) Vom Konflikt zur Lösung. Mentis, Münster (in press)
Shotter J, Gergen KJ (eds) (1989) Texts of identity. Sage, London
Sprung CL, Maia P, Bulow HH et al (2007) The importance of religious affiliation and culture on end-of-life decisions in European intensive care units. J Intensive Care Med 33(10):1732–1739
Stirling A (2008) "Opening up" and "closing down": power, participation, and pluralism in the social appraisal of technology. Sci Technol Human Values 33(2):262–294
Tao J (ed) (2002) Cross-cultural perspectives on the (im)possibility of global bioethics. Kluwer, Dordrecht
Van der Heide A, Deliens L, Faisst K et al (2003) End-of-life decision-making in six European countries: descriptive study. Lancet 362(9381):345–350 (on behalf of the EURELD consortium)
van Thiel GJMW, van Delden J (2010) Reflective equilibrium as a normative empirical model. Ethical Perspect 17:183–202
Van Tongeren P (1996) Ethics, tradition and hermeneutics. Ethical Perspect 3:175–184
Wahrman MZ (2002) Brave new Judaism: when science and scripture collide. Brandeis University Press, Hanover
Walter J (1999) Theological issues in genetics. Theol Stud 60(1):124–134
Young IM (2002) Inclusion and democracy. Oxford University Press, Oxford

MIX
Papier aus verantwortungsvollen Quellen
Paper from responsible sources
FSC® C105338

If you have any concerns about our products,
you can contact us on
ProductSafety@springernature.com

In case Publisher is established outside the EU,
the EU authorized representative is:
Springer Nature Customer Service Center GmbH
Europaplatz 3, 69115 Heidelberg, Germany

Printed by Libri Plureos GmbH
in Hamburg, Germany